BestMasters

Mit „BestMasters" zeichnet Springer die besten Masterarbeiten aus, die an renommierten Hochschulen in Deutschland, Österreich und der Schweiz entstanden sind. Die mit Höchstnote ausgezeichneten Arbeiten wurden durch Gutachter zur Veröffentlichung empfohlen und behandeln aktuelle Themen aus unterschiedlichen Fachgebieten der Naturwissenschaften, Psychologie, Technik und Wirtschaftswissenschaften. Die Reihe wendet sich an Praktiker und Wissenschaftler gleichermaßen und soll insbesondere auch Nachwuchswissenschaftlern Orientierung geben.

Springer awards "BestMasters" to the best master's theses which have been completed at renowned Universities in Germany, Austria, and Switzerland. The studies received highest marks and were recommended for publication by supervisors. They address current issues from various fields of research in natural sciences, psychology, technology, and economics. The series addresses practitioners as well as scientists and, in particular, offers guidance for early stage researchers.

Yvonne Plate

Social Identity Motivators in Environmental Collective Action

Patterns in Deciding to Participate in Extinction Rebellion

 Springer VS

Yvonne Plate
Tübingen, Germany

ISSN 2625-3577   ISSN 2625-3615 (electronic)
BestMasters
ISBN 978-3-658-44046-6   ISBN 978-3-658-44047-3 (eBook)
https://doi.org/10.1007/978-3-658-44047-3

This Springer VS imprint is published by the registered company Springer Fachmedien Wiesbaden
GmbH, part of Springer Nature.
The registered company address is: Abraham-Lincoln-Str. 46, 65189 Wiesbaden, Germany

Paper in this product is recyclable.

Abstract

Social Identity research has found prominence in the realm of collective action but lacks an environmental activism focus. Social engagement through collective action can influence political agendas. With the prominence of the climate crisis, it is thus relevant to investigate this realm further. This study gathers individuals' experiences to explore the influences on decision-making processes to join an environmental activist group—the local Extinction Rebellion (XR) group in Stuttgart, Germany. This case study is used to identify patterns in this process, to test the applicability of existing models—the social identity model of collective action (SIMCA), the encapsulation model of social identity in collective action (EMSICA), and the social identity model of pro-environmental action (SIMPEA)—and to explore the significance of social identity. The theoretical background gives an insight into some of the most prominent research on the topic to date, from the social identity approach to specific influencing factors. It emphasizes the importance of injustice perceptions, efficacy beliefs, norms and morals, and (politicized) social identity, as factors influencing collective action. A triangulation of methods—narrative interviews and observations—serves to explore individuals' experiences with relevant context. The interviewees are all activists or past activists, some involved in XR, and others in Fridays for Future. The results make it apparent that the existing models are not sufficient to represent social identity processes in environmental activism. The findings are visualized in a suggested adjusted model of collective action, which suggests norms and morals and (politicized) social identity as a twin core influencing collective action. Social identity retains its significance at the center of the model. Furthermore, it is suggested that collective efficacy beliefs and a sense of agency are exchangeable factors influencing the twin core and collective action directly. Injustice perceptions—acting as an umbrella term, also referring to environmental appraisals or moral convictions—similarly remain essential. Possible additional factors such as social connections, group identification, group dynamics, participatory efficacy, self-identity, and image are found to play a role occasionally but require further investigation. Further research into the environmental activism realm is encouraged, with the convergence of collective efficacy beliefs and a sense of agency, as well as the valuation of various factors presenting especially relevant topics.

Keywords: Social identity · Environmental activism · Collective action · Efficacy beliefs · Injustice perceptions · Norms and morals · Extinction Rebellion

Contents

Abbreviations

CDU	Christian Democratic Union
EMSICA	Encapsulated model of social identity in collective action
FFF	Fridays for Future
IPCC	Intergovernmental Panel on Climate Change
PCI	Politicized collective identity
PEB	Pro-environmental behavior
S21	Stuttgart 21
SCT	Social-categorization theory
SIA	Social identity approach
SIMCA	Social identity model of collective action
SIMPEA	Social identity model of pro-environmental action
SIT	Social identity theory
XR	Extinction Rebellion

List of Figures

List of Tables

Introduction

The climate crisis is a collective issue and therefore requires collective solutions. Environmental activism is one manifestation of humans working together in an attempt to communicate the gravity of the crisis, address the underlying systematic issues and prompt adequate solutions. The climate crisis presents a unique issue, due to several aspects. It is transnational and emerges at the intersection of human processes and the environment—geography, climate, and biodiversity. Its causes and effects are generally separated by time and space. Furthermore, there is no clear opponent or cause that can be fought but rather it is an interconnected structural issue with many instigators. This has made collective action groups addressing the climate crisis fundamentally different from historical collective action groups, beginning at their formation. The IPCC (Intergovernmental Panel on Climate Change) report, published in 2018, warns of dire consequences of global warming past 1.5 °C, and acknowledges the challenges faced in meeting that target:

> "Pathways limiting global warming to 1.5°C with no or limited overshoot would require rapid and far-reaching transitions in energy, land, urban and infrastructure (including transport and buildings), and industrial systems (high confidence)" (IPCC, 2018, p. 15).

This report reinvigorated the climate movement and birthed groups like *Fridays for Future* (FFF) and *Extinction Rebellion* (XR), with the goal of convincing governments that fast action is necessary. Many groups have expanded this goal to include more systematic and social justice issues, which can encourage commitment to stay with action groups (Fisher, 2016). Looking at the issue as being interconnected with other social issues not only gives a realistic image of the scale of the problem but could also strengthen a sense of urgency.

© The Author(s), under exclusive license to Springer Fachmedien Wiesbaden GmbH, part of Springer Nature 2024
Y. Plate, *Social Identity Motivators in Environmental Collective Action*, BestMasters, https://doi.org/10.1007/978-3-658-44047-3_1

Collective action is necessary specifically in climate change, due to its collective nature. It is not solvable by individuals and presents a commons dilemma (Hardin, 1968), as is often used to describe incidental environmental issues. Such a dilemma, where rational individuals consulting their own self-interest would deplete a limited shared resource (in this case the Earth and its resources), requires communication and collaborative solutions to benefit everybody long-term. While the ways in which to act against climate change are often a source of dispute, the influence of collective climate advocacy on policy is indisputable. Protest is a means for the public to get involved with politics. It can bring vital issues—such as the climate crisis, as specified in the IPCC report—to the forefront of the political agenda. The collaboration between activism and the scientific community is effective in bringing necessary issues to the political agenda (Thackeray et al., 2020). This collaboration through collective action is vital. Thus, it becomes pertinent to explore how individuals come to the decision to join collective actions to solve this collective issue.

As previous studies have mainly focused on social movements that did not address such complex issues, this realm allows for further exploration. The social aspects of the decision-making process are the focus of this research, due to the collective nature of the action and action groups. As a case study, XR acts as an example of a group engaging in clear and strong actions and additionally provides a perspective of a group that sets itself apart with non-hierarchical structures, non-violent civil disobedience, and scientific prioritization. It initiates further research into slightly more radical environmental movements (in comparison to e.g., FFF) and how groups form in this realm, of which there has been little research.

This study argues that the research on social identity in collective action is insufficient in addressing environmental collective action groups and the influences on individuals' identification with and participation in those groups. This thesis will use narrative interviews, observations, and deductive coding to address the following research questions:

- What patterns can be identified in the decision-making process of people joining the environmental activist group Extinction Rebellion in Stuttgart, Germany?
- To what extent are the existent models SIMCA, EMSICA, and SIMPEA applicable to this environmental activist realm?
- How does the significance of social identity manifest itself in the decision-making process?

The research is based on previous social identity research, including but not limited to the social identity model of collective action (SIMCA; van Zomeren et al., 2008), the encapsulation model of social identity in collective action (EMSICA; Thomas et al., 2009) and the social identity model of pro-environmental action (SIMPEA; Fritsche et al., 2018). As such, XR has not been investigated in this realm. The focus will not be on the belief in climate change, but rather the decision of how to act on it.

The structure of the thesis is as follows: Chapter 2 will provide a background on XR and their relevance as a social movement. The theoretical background will be laid out in Chapter 3, introducing the three models SIMCA, EMSICA, and SIMPEA and expanding on their factors and additional ones that influence collective action—subchapters *Identity, Norms and Morals, Efficacy,* and *Emotions*. The methodology in Chapter 4 will explain the research aim, the specifics of the methods, and their relevance. Chapter 5 will present the results of the research, introducing the interviewees and the coding scheme and covering the detailed coding analysis. The discussion in Chapter 6 brings together the previous chapters and discusses the research questions, ending with a consideration of possible limitations and future research opportunities. Conclusions are drawn in the final chapter, Chapter 7.

Background: Social Movements, Climate Justice, & Extinction Rebellion

Social movements play an imperative role in the world today. From engaging in democracies to addressing crucial issues, they serve many purposes. The communication of collective opinions in such present and disruptive ways allows movements to set political agendas, influence decision-making and make demands of what they feel is deserved (Amenta et al., 2010). Global climate movements using civil disobedience in combination with scientific backing are suggested to increase public awareness and the public's engagement with climate issues (Thackeray et al., 2020). Activism in the form of environmental movements is necessary to ensure the serious consideration of scientifically proven issues. As Thackeray et al. phrase:

> "Science without activism is powerless to enact change, but activism without science will enact change without knowledge of the direction in which change is needed. To make constructive progress, both science and activism are needed to move society in the right direction with strength and purpose" (Thackeray et al., 2020, p. 1043).

The collaboration of these two groups—the scientific community and social movements—brings constructive progress to society (Thackeray et al., 2020). This method of bringing forward issues also allows for more diverse voices than might be present in the government to demand a say in their future.

While it might not be possible for governments to prioritize all issues, social movements bring to light the importance of diverse issues that might have been neglected. As Bell and Bevan (2021) highlight, this also requires the involvement of a diverse set of perspectives in the environmental movement, pointing towards the *Standpoint Theory*—first developed in feminist theory by Harding (1986)—which implies that diverse groups need to be represented to ensure the provision of important perspectives from their experiences. The representation of a diverse

© The Author(s), under exclusive license to Springer Fachmedien Wiesbaden GmbH, part of Springer Nature 2024
Y. Plate, *Social Identity Motivators in Environmental Collective Action*,
BestMasters, https://doi.org/10.1007/978-3-658-44047-3_2

set of groups both in different movements and within movements brings forth perspectives that otherwise fall under the radar or are overlooked completely. In environmental issues, this can be especially relevant due to the disparity between the causers of the issues and the affected. This disparity is an *Environmental Justice* issue where environmental and social injustices intertwine (Walker, 2020). The societal aspect of environmental issues is integral to finding fair and just solutions, which is why the diversity of groups involved in bringing to light issues is crucial. Climate justice is intrinsically linked with environmental justice which both highlight the fact that environmental and climate issues are unequally created and their effects unjustly distributed. It is a mantra[1] often heard in climate marches and as such it is a common aspect prioritized in many environmental movements. This necessitates diverse movements and movement participation to achieve true justice (Schlosberg & Collins, 2014). So, diversity is crucial in the movement itself to ensure a just movement, as much as it is crucial in the variety of social movements to bring forward diverse issues.

The year 2018 showed a resurgence of the climate movement with both Fridays for Future and Extinction Rebellion, as well as other environmental groups emerging and gaining momentum. This resurgence coincided with the IPCC report that communicated that only 12 years were left to support radical changes to avoid the worst effects of the climate crisis (IPCC, 2018; Taylor, 2020). The groups' main goal was and still largely is the communication of this call for action from the science community (e.g., Extinction Rebellion, n.d.). They fight to bring this neglected issue to the forefront of the political agenda. The environmental groups in this movement are far-reaching and diverse and they bring forth various local and international issues (Weyler, 2018), with climate justice well represented in the discourse. The movement thus fulfills the important purpose of bringing constructive progress through the collaboration of science and diverse activism, as far as it succeeds.

XR emerged towards the end of 2018 in the United Kingdom and shortly thereafter spread internationally. It is a more radical group that fights for governmental action on mass extinction, climate change, and social and ecological collapse, citing their focus on basing action on scientific evidence. They describe themselves as follows:

---

[1] "What do we want?", "Climate Justice!", "When do we want it?", "Now!" (Balthesen, 2019)

"Extinction Rebellion is a decentralized, international and politically non-partisan movement using non-violent direct action and civil disobedience to persuade governments to act justly on the Climate and Ecological Emergency" (Extinction Rebellion, n.d.).

The main goal is to do more than march for justice and action on climate and environmental issues. Civil disobedience can be defined as "a public, nonviolent, conscientious yet political act contrary to law usually done with the aim of bringing about a change in the law or policies of the government" (Rawls, 1971, p. 364). Though often controversial and involving the risk of arrest, this method has found success. This success is what motivated the group in the first place (Taylor, 2020). The strength they saw in the method and their desire for change motivated their collective action. As it spread into over 1000 local groups around the world, the group developed 10 principles and values such as a shared vision, a goal of engaging 3.5% of the population, breaking down hierarchies, and avoiding shaming and blaming (Extinction Rebellion Germany, n.d.). These are required to be followed by all local groups as they plan their own actions (Taylor, 2020), to remain coherent and united under the common XR banner. Their three demands (Extinction Rebellion, n.d.) also remain largely the same internationally:

1. *Tell the truth.* Governments must declare a climate and ecological emergency and communicate urgency.
2. *Act now.* Governments must act now to halt biodiversity loss and reduce greenhouse gas emissions to net zero by 2025.
3. *Go beyond politics.* Governments must create and be led by the decisions of a Citizens' Assembly on climate and ecological justice.

XR's more radical approach has been met with criticism, despite their prioritization of science and emphasis on nonviolence. This has led to some seeing it as a necessary group, but others seeing it as extreme (Taylor, 2020). The activist tactics involving arrest were specifically criticized on account of building on white privilege (Bell & Bevan, 2021) and overlooking the issues of police racism (Taylor, 2020), which make civil disobedience a much more accessible tactic for white people. These criticisms might point to a feeling of exclusion or alienation of less privileged people from XR, which may even lead them to reject environmentalism more generally (Bell & Bevan, 2021). Additionally, although one of XR's goals is diversity, also politically, for example in not only involving leftists (Fahrion, 2019), criticisms have surfaced claiming a lack thereof. Their aim for inclusion and a decentralized model has not always brought about true diversity, with societal structures prevailing within the realms of the group. There are claims that the more privileged, often white men, take over, which undermines the aim for diversity (Taylor, 2020). Though these issues and criticisms were more prevalently discussed in the UK, it is a vital discussion to keep in mind with the importance of diversity in environmental issues. It is something that the group is actively working on (Extinction Rebellion, n.d.), but it could also be a noteworthy factor to investigate when looking at the decision to join XR.

In Germany, the first local XR group was created at the end of 2018 and less than a year later 50 separate XR groups had been established throughout the country (Fahrion, 2019). To prepare newly recruited members around Germany for actions, the first phase involved various lectures, training, and tips that taught them how to safely join actions and effectively participate (Extinction Rebellion Germany, n.d.). The first actions started in the spring of 2019 with two bridges and important transportation passages being occupied, one in Berlin (Oberbaumbrücke) in April and one in Cologne (Deutzer Brücke) in July (Fahrion, 2019). Further street blockages and other occupations of public space were possible for the new German groups before the COVID-19 pandemic spoiled plans for further actions in 2020 (Extinction Rebellion Germany, n.d.). This slowed down the momentum built up before the pandemic, as activist actions were shifted online, largely impacting groups like XR. The events of the year made possibilities more tangible, watching the prominent *Black Lives Matter* movement at the forefront of the public discussion and large-scale protests all around the world. At the same time, climate concerns were overshadowed by the pandemic and there were only limited possibilities for public action. Nevertheless, the following year brought about new energy with action weeks, blockages, protests, and other creative and symbolic actions happening around the country (Taylor, 2020). With their activity in Germany, XR has experienced some successes for their third demand—the establishment of a Citizen's Assembly—for example in Berlin, though not binding by law (Extinction Rebellion Germany, n.d.). More recently this has also become a tangible reality for the local group in Stuttgart—the focus of this research—as will be seen in the results of this study.

The local XR group in Stuttgart was a part of the first wave of local action groups formed in 2018. Outside of street blockages and involvement in marches, the Stuttgart XR group also initiated some unique and regionally relevant actions (Extinction Rebellion Germany, n.d.). The largest ones documented on their website are a protest against the Stuttgart 21 (S21) train station project and a call out of the political parties' climate policies before the state elections. The S21 train station project—a large rebuild of the Stuttgart main train station—is a controversial project that has gone over the planned time, budget, and environmental harm (e.g., the amount of environmentally harmful concrete used). XR criticizes this, as they see the project as unnecessary. As the undertaking is too far along in construction to be abandoned entirely, the protest advocated for an alternative direction to be taken from this point forward, which would be more in line with climate politics. They also demanded that the current environmental impacts be tested. To emphasize their point, the action involved a couple of XR participants climbing onto a crane on the construction site to occupy it for a night and hang up a banner with the message *"S21 = KLIMA KILLER"*, calling the project a *'climate killer'*. The banner also featured the XR logo and the logo for the alternative plan *umstieg-21*, which translates to *'changeover-21'*. Police were called in response to this action and the S21 construction is still ongoing. For the state elections (Landtagswahl) in Baden-Württemberg in 2021, the Stuttgart XR group attempted to hold the parties accountable for the climate goals. They accused the green party Bündnis 90/Die Grünen of prioritizing the car industry over climate concerns despite their self-proclaimed green focus, and the conservative party CDU (Christian Democratic Union), which has been involved in the state government for several decades for not taking climate issues seriously. Both parties received letters from XR making clear their disappointment and demands. Additionally, XR created fake CDU election posters with more 'real' mottos such as *"CDU, weil es auf einem toten Planeten noch Jobs geben muss."*, which translates to *'CDU because there must still be jobs on a dead planet.'*. These posters were hung up around cities state-wide. XR Stuttgart also lists actions like food-saving in solidarity with *Letzte Generation* (a more radical group) and a street blockage action to advocate for a car-free downtown on their website (Extinction Rebellion Germany, n.d.). Their social media pages also make clear their involvement with *Lützi bleibt ('Lützi stays')*, a country-wide action (with international involvement), very present in the media, where activists occupied a German town near the Dutch border that was planned to be torn down for brown coal mining. These actions and associations give an impression of both the variety of actions they have organized and been involved in, as well as some of the publicity they might have created in the area. Outside of actions, they have also organized various community events like a vegan barbeque, resilience workshops, walks, and evening events with lectures and music. So, while their most prevalent presence towards the public is very expressive actions and demands, they also put a focus on community building and taking care of each other.

Theoretical Framework 3

Motivations for pro-environmental action can be considered through both individual and social lenses. For collective action, though, social and group connections are more significant than motivators for individual behavioral change (Schmitt et al., 2019). Action is defined as 'collective' whenever "disadvantaged group members self-categorize as a member of that group and are motivated to achieve that group's goals (e.g., to improve the group's conditions or more generally to seek social change) through some form of action (i.e., signing a petition, attending a mass demonstration)" (van Zomeren, 2016, p. 90). Collective action in this context presupposes the need for a social group. The individuals need to identify as part of a group and act in its interest to participate in collective action. This chapter will introduce some of the numerous theoretical approaches to collective action and what motivates it. Arguably the most common consideration in collective action is this social nature and the need for social identity. Therefore, the first subsection will summarize the social identity approach, which the models introduced in the subsequent sections build upon. These models consider the connections between predictive factors and social identities in relation to collective action. The applicability of these models will be considered throughout the thesis, as well as possible points of expansion or adjustment. As the models only give a limited framework on motivators for collective action, the following theory subsections on identity, norms, efficacy, and emotions will cover additional theory and proposals for influences on participation. Not all introduced theory and terms will directly aid the current study but rather are intended to add to a holistic overview of existent research.

Y. Plate, *Social Identity Motivators in Environmental Collective Action*, BestMasters, https://doi.org/10.1007/978-3-658-44047-3_3

3.1   The Social Identity Approach

The social identity approach (Reicher et al., 2010) is integral to the consideration of environmental collective action. The environmental crisis presents a global problem caused by a plethora of actors and cannot be fought on an individual level. Therefore, the social capabilities of humans are necessary in order to work together to combat the consequences and further deterioration of climate and environment crises. The social identity approach (Reicher et al., 2010) consisting of social identity theory (SIT; Tajfel, 1978) and self-categorization theory (SCT; Turner et al., 1987) gives insights into how individuals become part of a group and how they interact with and within such a group. This approach has been used as a foundation for a variety of collective action research (e.g., Fritsche et al., 2018; Thomas et al., 2009; van Zomeren et al., 2008). SIT defines the social self as "that part of an individual's self-concept which derives from his knowledge of his membership in a social group (or groups) together with the value and emotional significance attached to that group membership" (Tajfel, 1978, p. 63). In other words, social identity is the way in which a group and the group's significance to the individual defines and contributes to their identity. It connects individuals to the social world, as it is relational, shared, and meaningful. It is relational, as it is an individual defining themselves in terms of their similarities and differences compared to others. It is shared with others, which allows for shared action. And its meanings are derived from a collective past and present (Reicher et al., 2010). SCT complements this with a proposal on how individuals, through this social identity, are able to act as a group. Part of this is linked closely to SIT, as it sees the shared social identity in contrast to others and as a basis on which attitudes and behaviors are based. SCT further proposes individuals, as part of a group, are able to depersonalize their individual self-perception and behavior in favor of internalizing the social whole (Turner et al., 1987). This level of abstraction allows them to act as representatives of the group and for the group to act as a whole. SCT thereby emphasizes the potential of collective psychology, as the social identity becomes an integral part of the self. SIA can give practical insights into group dynamics and how they fit into the social world, transferring its use from theory to a new scope (Reicher et al., 2010). This ability to become a part of a group and act with a shared social identity is fundamental to understanding collective action and how individuals are able to become a part of a movement. As such, very impactful models of collective action are based largely on the SIA.

3.2   The Social Identity Model of Collective Action

The social identity model of collective action (SIMCA; Figure 3.1) is a widely applied model, which incorporates the SIA to frame social identity as a defining factor in collective action (van Zomeren et al., 2008). It is based on a meta-analytical approach of the available collective action literature, integrating three significant socio-psychological perspectives[1]. The SIMCA predicts three factors from these socio-psychological perspectives, which also interact with each other, to affect collective action—social identity, injustice beliefs, and group efficacy beliefs, as seen in Figure 3.1. Arguing that people will act if they view an issue as unjust, they believe their group is effective in righting that injustice and they identify with a group that can mobilize collective action. Identity is posited centrally, as it motivates collective action both directly and indirectly through injustice and efficacy perceptions, inherently interlinking these factors. These three factors—peoples' senses of identity, efficacy, and injustice beliefs—are shown to causally affect and predict attitudes, intentions, and behaviors in collective action. The SIMCA follows classic psychological attitude-behavior models, where behavior shows a smaller effect than attitude, as it is not idealistic. Behavior is bridged through intentions and is affected by not only practical limitations but also random or systematic factors (van Zomeren et al., 2008). Several studies have supported the SIMCA or updated versions of it internationally and pertaining to different issues, even receiving some support in the environmental youth movement (van Zomeren et al., 2018). Additionally, outside of this model, other studies (e.g., Xiang et al., 2019) continue to find virtually the same interrelations of factors as the SIMCA. In positioning social identity so centrally, its primary purpose, providing agency that would otherwise not be possible, is made apparent. The collective action in this model remains dependent on the content of that social identity. This social identity acts as an ingroup, which is treated unjustly by an outgroup (or an other). To motivate action, individuals that are a part of this ingroup must perceive it as being effective in righting this observed injustice. The pathways that are established in SIMCA, where social identity bridges the explanations of collective action for injustice and efficacy, were found to be the best empirically fitting model (van Zomeren et al., 2008). These indirect paths in which identity affects collective action are further addressed and explained through group-based processes involving emotions and efficacy.

---

[1] See van Zomeren et al. (2008) for elaboration on perspectives (identity approach, group efficacy approach and injustice approach).

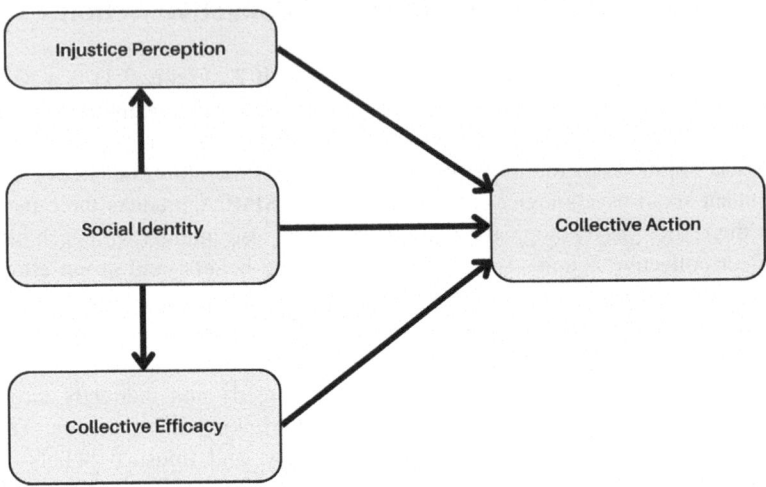

Figure 3.1   Original SIMCA (van Zomeren et al., 2008)

Social identity informs group-based emotions (e.g., perceptions of injustice) and group efficacy. Therefore, according to the SIMCA, additionally to its direct motivation, it is also a significant indirect motivator of collective action. As social identity provides the basis for emotion, action, and perception in the group, it predicts injustice felt by the group and thus the individual, as well as group efficacy as felt by the group. By explaining both injustice and efficacy as motivators through identity, the factors are integrated harmoniously. Nevertheless, identity, efficacy, and injustice must be considered separately in this dynamic model of interactions. The individual factors are distinguished further by van Zomeren et al. (2008).

The SIMCA shows a stronger effect of affective rather than non-affective injustice, as well as politicized identity over non-politicized identity (van Zomeren et al. 2008), which more recent studies build on further (as will be covered in the upcoming sections). Affective injustice here is the feeling of receiving unjust treatment, whereas non-affective injustice refers to recognition and perceptions of injustice towards others. The SIMCA directly connects affective injustice to group-based emotions, perceptions, and actions originating from social identification. A group is more likely to act if it perceives itself as receiving unjust treatment. Politicized identity is also a significant factor, as it predisposes a group to jump to action more quickly. Van Zomeren et al. (2008) incorporate the significance of the politicized identity from Stürmer and Simon (2004), who argue that people can experience a stronger inner obligation to act when their social identity is inherently political, as, for example, with a social movement. This political social identity is in that way often already tied to collective action, which reduces the barrier to action. If a group is defined by action, it is likely to engage in it.

Van Zomeren et al. (2008) further differentiate the effect of injustice and efficacy perceptions in the type of disadvantage that is being addressed. The sense of injustice and efficacy beliefs affect collective action more in issue- or situation-based disadvantages—incidental disadvantage—than in disadvantages that are structurally founded—structural disadvantage. The social identity factor does not play into this differentiation, as it is present in and thus predicts both disadvantages. Incidental disadvantage usually requires a social identity to be formed around the situation, which can be a barrier but can also make it easier to incite collective action, as that is the purpose of the newly formed group. Structural disadvantage contrastingly often presupposes a social identity that is embedded historically or socio-structurally. This can make it more of a challenge to incite action, as the disadvantage can be much more ingrained and accepted in the self-identity. As such, injustice perceptions could be hindered. Additionally, it is usually a smaller task to fight against an incident rather than a much bigger, more powerful system. The more novel situation of incidental disadvantage, as well as the less embedded nature, allows for a larger effect of a sense of injustice on collective action. It is propositioned that group-based emotions originate from subjectively assessed situations. Thus, incidental disadvantages more strongly than structural ones can evoke feelings of injustice, as there is likely a more present situation to assess. This mechanism does not have to be limited to feelings of injustice but can escalate to group-based emotions of anger originating from these feelings of injustice. These more easily incited group-based emotions can thereby more directly lead to collective action (van Zomeren et al., 2008). Since the SIMCA has differing effects depending on the type of disadvantage, it can be critical to distinguish between them. The SIMCA did not stop at this first iteration, rather, it was further extended as the extended SIMCA (Figure 3.2).

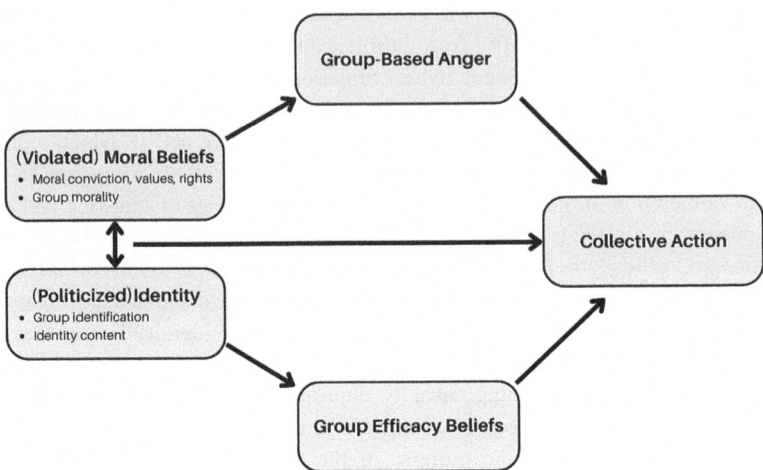

Figure 3.2   Extended SIMCA (van Zomeren et al., 2018)

A more recent revision of the SIMCA lays a greater focus on moral beliefs and their violation, as well as identity and its politicization through a normative fit model based on SCT (van Zomeren et al., 2018). The relations stay largely the same, but the revised model posits moral beliefs in a similar position to social identity. It suggests that the normative fit categorization—social categorization based on norms—in response to violated moral beliefs, should be viewed separately but intertwined with the (politicized) identity. Violated moral beliefs serve the same role as perceived injustice, where an issue is deemed unjust or immoral and needs to be rectified. The politicized identity in this model is both normative and action-oriented and surpasses group identification in importance. The content of the identity is set above the identification itself. Politicized identity only acts effectively with action-oriented identity content. Moral and identity motivations are posed as the "twin core" of the revised SIMCA, as a person's identity and what they stand for are intrinsically intertwined (van Zomeren et al. 2018, pp. 124–125). Van Zomeren et al. (2018) set the moral categorizations separate from opinion-based groups (Bliuc et al., 2007), which are covered in the coming *Identity* subchapter. In the extended SIMCA, the social identities are moralized and politicized, rather than purely opinion based. It suggests the significance of these strong identities is people protecting who they are, and therefore being strongly motivated to act. Through the promotion of moral beliefs, this extension of the SIMCA adds another dimension to the motivation for collective action, which will be taken up in the *Norms and Morals* subchapter.

3.3 The Encapsulation Model of Social Identity in Collective Action

The encapsulation model of social identity in collective action (EMSICA; Thomas et al., 2009; Figure 3.3) provides a similar approach and model to the original SIMCA with the same set of factors predicting collective action but proposing the reversal of the indirect effects. Similarly to the SIMCA, the foundation of this model is the SIA. Thomas et al. (2009) likewise promote the idea that social identities (associated with the relevant norms for the required action) are a key motivator of collective action. However, in the EMSICA the identity factor acts as a mediator. The way in which the factors are ordered in this model, perception of injustice, and efficacy beliefs inform the collective self, rather than originating from it (Thomas et al., 2009). The individual takes into consideration if their own emotions align with the group's and if they believe this group is efficacious, before identifying themselves with it. Through this process and the acknowledgment that multiple complex processes are possible, the EMSICA places a focus on the creation of group cohesion and identity without losing sight of the individual.

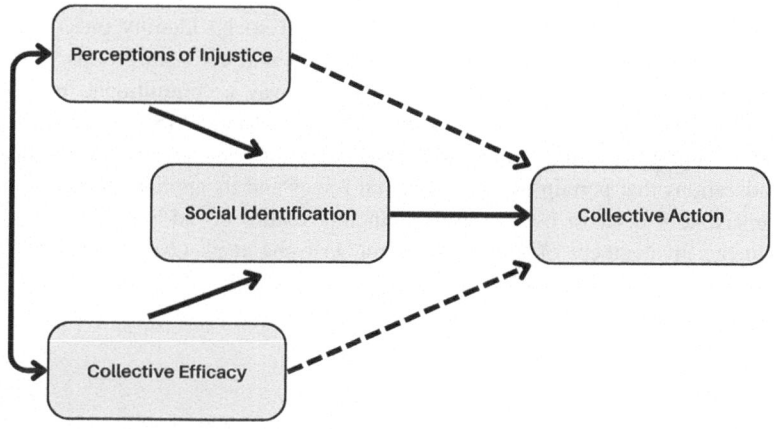

Figure 3.3 EMSICA (Thomas et al., 2009)

Thomas et al. (2012) have found support for both SIMCA and EMSICA, explaining that shared emotions and beliefs can be encapsulated in identity-forming processes, but correspondingly identity-forming processes can also shape and create shared emotions and beliefs. Thus, it can be seen as a challenge to distinguish causality. These seemingly contradictory causal relations can also point to a more complex, dynamic system of interaction in which identity processes come in different forms and norms may play into them more significantly as well. Thomas et al. (2009)—simultaneously with the EMSICA—suggest a complimentary model of normative alignment, which can incorporate these new considerations and will be elaborated on in the subsection *Norms and Morals*.

While SIMCA provides a good foundation for the research of collective action in a social realm, the environmental movement presents a more complex issue than most other challenges addressed by collective action. It is an issue that affects everyone but often not directly. So, rather than an injustice towards a specific group of people (though it can involve this), the environmental movement is about the fight against injustice towards the earth and humanity as a whole. It is a complex combination of both incidental disadvantages as well as much more structural issues. The latter spawn the situational issues, as well as much more all-consuming disadvantages for humanity. EMSICA similarly fails to acknowledge this uniqueness of the environmental movement in its function. Furthermore, environmental movement groups are groups in response to the climate crisis and do not exist outside of that issue. Therefore, SIMCA's relational structure fails to address how these groups come together and what could motivate individuals to join. Rather, they focus on the influence of social identity on collective action, which is already presupposed in environmental activism groups. To join an environmental activism group is already in a way a commitment to action. To take this step is to agree with an injustice. Therefore, EMSICA seems to be the more fitting relational alignment. Nevertheless, it does not include important considerations that pertain to environmental issues and its applicability and comprehensiveness need to be investigated in this realm. To address this research gap in the involvement of the environment, Fritsche et al. (2018) developed an altered model, taking into consideration pro-environmental behavior in response to large-scale environmental crises.

3.4   Social Identity Model of Pro-Environmental Action

The social identity model of pro-environmental action (SIMPEA; Fritsche et al., 2018; Figure 3.4) builds on the SIMCA and the EMSICA in an environmental behavior dimension. It illustrates the vital differences in the environmental realm but places its focus on general pro-environmental behavior (PEB), defined by Stern (2000) to include environmental activism, non-activist behaviors in the public sphere, private sphere environmentalism, and other environmentally significant behaviors. By building on previous collective action models, Fritsche et al. (2018) consider the pro-environmental behavior in its collective dimension to propose a model specifically for this topic. The connection the authors make apparent is that this new model is necessary to capture the collective nature of a lot of perception shaping and group inclinations in environmental behavior and goals, as well as the need for collective action to successfully work towards said goals. Large-scale environmental crises are often subject to existing more abstractly for individuals, as there is usually a gap in time and space between who caused them and who will suffer from them. This means environmental appraisals and behavior, according to Fritsche et al. (2018), are largely determined by collective thinking and processes to bridge this gap.

The SIMCA and the EMSICA are not sufficient to explain PEB for several reasons. The disadvantages, as seen in these models, are not quite comparable to the environmental issues and the global climate catastrophe that similarly need to be fought by collective action, as explained previously. Additionally, the lack of a clear other in climate change and many other environmental issues sets the environmental movement apart from other social movements. It can be much more challenging to fight issues when there is no clear opponent (Fritsche et al., 2018). This makes the SIMPEA a valuable contribution to the collective action research and relevant to look at for current research as well. However, one of the authors' other main reasons for this new model is the inclusion of not just collective action, but also individual environmental behaviors. Fritsche et al. (2018) point out the fact that emotions and motivations of any environmental behavior (not solely activism) can be driven by collective as well as personal emotions and identity drivers. This makes this model very valuable but more convoluted than necessary for the current research. However, while the broader frame does not fit perfectly when looking at activism specifically, the model is still valuable for the above reasons, as well as its inclusion of norms.

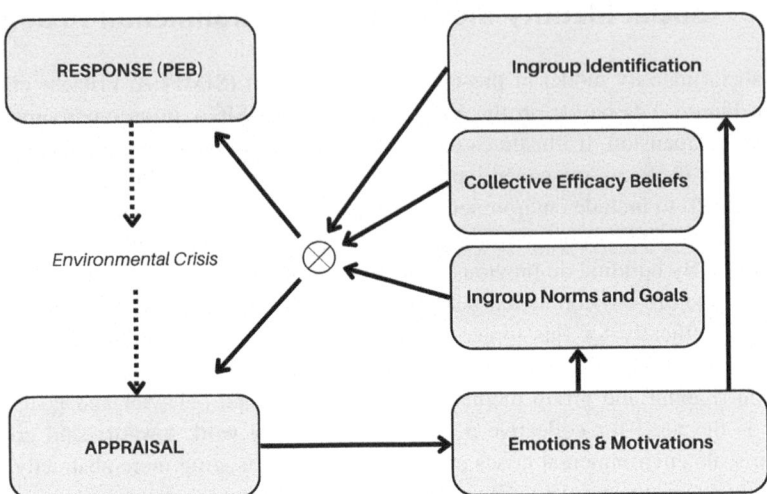

Figure 3.4 SIMPEA (Fritsche et al., 2018)

The SIMPEA (Fig. 3.4) promotes ingroup norms and goals as one of the main three drivers of action and appraisal—the other two being ingroup identification and collective efficacy. Collective and personal motivations and emotions in this model become an indirect effect, but still play a role in pro-environmental action motivation as they arise from environmental appraisals and thus motivate the previously mentioned drivers. The integration of norms as a significant factor, as mentioned in previous sections, has been a common finding and its interplay here with emotions and motivations is a noteworthy relationship to consider. This proposed model is largely based on previous studies and experimental evidence, though there is still a lack of backing on the interactive effects between the three main motivators of PEB (Fritsche et al., 2018). With the SIMPEA, Fritsche et al. (2018) propose that a perceived environmental threat or threat to social identity motives (through environmental appraisal) intrinsically incites collective responses to environmental issues, which can increase individuals' involvement in groups and collective action, through the proposed drivers. Rather than intergroup conflict and resulting anger as a group-based emotion, as proposed in the SIMCA (van Zomeren et al., 2008), in the SIMPEA it is this environmental threat that is enough to motivate collective action (Fritsche et al., 2018). This is a pivotal factor in furthering previous models for environmental crises. The initiating motivator can be fundamentally different, as the other as such is often not apparent.

The perceived injustice—represented here by environmental threat perception—does not come from a specific instigator but is rather a much more complex result of faulty social systems. The perceived environmental threat becomes the determining factor instead, which materializes the faulty social system in a more specific issue, which can be perceived as unjust or immoral. The complexity of these environmental crises remains essential in the consideration of collective environmental action. Nevertheless, this model has its shortcomings, as it does not focus solely on activism and does not place the social identity centrally. This is necessary for stronger forms of activism, as individuals mainly engage in this form of collective action through activist groups, as training and planning is required. As the presented models give a concise but somewhat limited view of environmental collective action, the coming sections will elaborate on more recent developments on driving factors, as well as significant specifications and contributions that may aid the understanding of collective action. Furthermore, these sections will aim to tie in the environmental and activism components in order to define the theory more concisely.

3.5   Identity

The most consistent and prominent factor: identity, as a rather broad concept can benefit from elaboration in its self and group components as well as in its politicization and opinion-grouping. Identity as a dominant theme has been consistent across all three models in the centrality of a collective identity, a social identity, or a group identity. Its significance is widely agreed upon far beyond the models presented above, with studies showing the strength of its direct effect on participation intention (e.g., Bamberg et al., 2015), as well as its significance in and interaction with other variables (e.g., motivations and emotions; Haugestad et al., 2021). This strength of the group identity to motivate action can be especially significant in environmental issues, where personal efforts can seem and be insignificant (Fritsche & Masson, 2021). The connection to a group can be defining in making individuals feel in control (Jugert et al., 2016). This feeling can be a vital motivation in the face of global climate issues which can quickly seem hopeless. The individual can only be empowered to act if they feel that they can gain some control over the issue. Communicating the efficacy of group action in combination with the consequences of climate change has been suggested to avoid this hopelessness upon learning about an issue (Jugert et al., 2016). By immediately promoting this group identification, instead of falling into hopelessness, individuals may feel more in control and thus be motivated to act. Fielding et al. (2008) also found specifically that social identity in the form of environmental group membership was a strong determining factor of individuals' intention to participate in environmental activism. This could relate to the previously mentioned feeling of being in control. When the group is seen as a collective agent, then this social identity can be operationalized into action (Fritsche & Masson, 2021). This form of the social self can make individuals capable and motivated to act collectively (Jugert et al., 2016). While this strong effect of group identity is a primary predictor of collective action, it can be worth looking at the way individuals identify themselves and how this influences their decisions and actions as well.

Self-identity, or the way individuals see themselves can additionally play a part in the motivation to join collective action. While its impact can be moderated by group membership (Fielding et al., 2008), the self-image can be a powerful motivator towards actions that support and build it and away from actions that would go against it. This may also play a part in historical participation in collective action, motivating further collective action participation (Bamberg et al., 2015), as the self-image of a participant is built up. In other words, in order to strengthen one's self-image, one would engage in behavior that supports it. Thereby, individuals that see themselves as environmental activists have also been found to have more intentions to engage in future activism (Fielding et al., 2008). This effect was found to be especially significant at low group participation, where group norms have little to no effect on the individual and was much less significant when there is high group participation, where group membership tends to moderate the effects and become a more significant predictor (Fielding et al., 2008). In the case of high group membership, the individual's self-image becomes more irrelevant in comparison to the group norms that drive activism intention. The collective identity becomes more influential than the self-identity. This group membership motivation is most present in groups that can target it in a specific direction, for example, opinion-based groups.

The EMSICA specifies that opinion-based groups provide a stronger social identity to promote social change (Thomas et al., 2009), while van Zomeren et al. (2018) argue that merely opinion-based groups do not go far enough and that moralized and politicized identities are the true drivers of collective action. The definition used refers to Simon and Klandermans:

"[…] people evince politicized collective identity to the extent that they engage as self-conscious group members in a power struggle on behalf of their group knowing that it is the more inclusive societal context in which this struggle has to be fought out" (Simon & Klandermans, 2001, p. 319).

Opinion-based groups are groups based on shared opinions and norms that relate to positive social change and can include both advantaged and disadvantaged people (Thomas et al., 2009). Membership in and identification with such a group is likely to motivate and sustain behavior in support of that social change. These psychologically meaningful groups are, therefore, argued to be a strong predictor of action intentions. Bliuc et al. (2007) supports this claim by suggesting that this self-definition with a *political* opinion-based group is a significant predictor in line with predictions from SCT. As can be seen from this phrasing and argumentation, the political aspect is not absent from opinion-based groups. Nevertheless, some studies argue for the importance of specifically politicized identities (e.g., van Zomeren et al., 2018). Further research, including more recent studies, that have evaluated the most effective social identities for collective action have zeroed in on politicized identities (Haugestad et al., 2021; Stürmer & Simon, 2004; van Zomeren et al., 2018), arguably just a more defined and insistent political opinion-based group that is actively engaged in a power struggle. The politicization argument even pre-dates the more specific models of collective action based on social identity. Simon and Klandermans (2001) present this identity as a model, the politicized collective identity (PCI), and made apparent its potential for agency. The PCI model suggests this political identity is based on three pillars—collective identity, power struggle, and societal context for said struggle—and to be developed through three stages—awareness of shared grievances, adversarial attribution, and involvement of society as a third party (in addition to the in- and outgroup) by triangulation—to be able to engage in a power struggle. This power struggle is fought by the self-conscious members on behalf of their identified group, against an outgroup, and with the awareness of its positioning in the wider societal context. While the outgroup, as argued before, does not exist as such in the environmental context and the PCI model does not specify collective action, the politicization of identities is still relevant to discuss in the case of activist groups. In this model, as PCI builds on collective identity, this collective connection is reinforced, and collective identity produces more intense effects (Simon & Klandermans, 2001). The politicization also ties back into the view of a group as a collective agent. As it is a goal-directed identity, individuals can gain a feeling of agency through membership and can through that perceive the group as more efficacious (Simon & Klandermans, 2001), which can strengthen identification, participation, and action. Such collective efficacy, as well as participative efficacy ties in seamlessly here and will be covered more in-depth in the coming *Efficacy* subsection. The PCI model makes evident how politicization can strengthen the effect of social identity on collective action. While a nonpolitical group can also be disadvantaged and decide to jointly act to fight for

justice, a politicized collective identity already pre-disposes the need and motivation for action. A social (and/or environmental movement) would classify as such
a politicized group and this politicization is placed on a level of high significance
in regards to activism (Haugestad et al., 2021; Schmitt et al., 2019; Stürmer &
Simon, 2004). Schmitt et al. (2019), who tested if activism followed the same
pattern as PEB with a large motivating factor of identification with nature, in
line with the previous points found that a politicized environmental identity is
actually a much stronger predictor and intermediary. This, as well as emphasizing the importance of politicization of identity, also reaffirms the need to look at
the social aspect when determining motivations for activism, as the patterns are
different than for other PEB.

3.6   Norms and Morals

The involvement of norms in the motivation of collective action has been a common theme throughout the previous sections. Norms, as well as moral obligations,
have been developed as an integral factor parallel and in relation to the presented
models. The normative alignment model (Thomas et al., 2009), developed along-
side EMSICA, is meant to provide the mechanism of encapsulation in group
formation (like the creation of consistent group characteristics and a cohesive
identity), to work in tandem with the EMSICA which solely places identity in
the mediation position. Thomas et al. (2009) suggest the permeation of norms
throughout all factors influencing action. Only the inclusion of these norms sup-
posedly makes motivation through identities possible in the first place. Therein, it
is suggested that a social identity with relevant norms concerning emotion, efficacy, and action is necessary to promote sustained collective action motivation.
The identities proposed to effectively fight the cause are ones where a politicized
group becomes associated with clear and relevant descriptive norms. Descriptive
norms here constitute what is actually done, while injunctive norms "specify the
rules or beliefs as to what constitutes moral and approved (or immoral and dis-
approved) behavior" (Thomas et al., 2009, p. 203). While descriptive norms are
represented in the present action, injunctive norms in this model can be what
drives stronger goals for the future. Aligning these two, especially in the context
of a politicized identity can be very effective in initiating behavior, as the progressive goals become the present actions. This alignment of norms also applies
to the other influencing factors. Thomas et al. (2009) underline the need for there
to be politicized (or as they call it 'pro-change') identities that believe they can
make an effective difference, feel outrage, and find it necessary and want to take

relevant action. These norms and their congruent alignment are all said to be necessary to sustain long-term success in social and political action. The normative alignment suggested is shown to be manufacturable through a process of opinion-based group interaction. Indeed, through an opinion-based group interaction method Thomas et al. (2009) found that bringing together individuals in a group with a goal to promote a common opinion was effective in strengthening the individuals' norms for action, efficacy, and emotion in support of the cause, though this was mediated by strengthened group identification. This model makes apparent the role norms can play and how other factors can interrelate.

Similar findings and conclusions have been drawn outside of this model that support the important role of norms in collective action. In the SIMPEA, for example, ingroup norms are also assumed to interact with collective identification and collective efficacy beliefs to predict behavior (Fritsche & Masson, 2021), showing a similar amount of integration as in the EMSICA, but as a separate factor. Normative support of collective activism has also been shown to increase intent to participate (Fielding et al., 2008). This supports the pervasion of norms throughout the interactions of group involvement and collective action. Similarly, Schmitt et al. (2019) found that politicized identity does not exist in motivating activism alone but is in collaboration with moral obligation, based on an individual's norms. So, if relevant norms are present, individuals will act on a certain moral obligation to act. This can fall in line with moral conviction, which will be picked up in the *Emotions* subsection. While the norms mentioned up to this point have considered individual norms and collective action group norms, normative involvement can also occur from other groups that individuals identify as members of. Rees and Bamberg (2014) found that perceived participation norms—the belief that it is expected by one's group or community to participate—were one of the strongest predictors of participation intention in a community initiative. It is, thus, vital to consider the individual as part of a larger system and not a stand-alone actor.

3.7 Efficacy

The significance of efficacy beliefs follows the reasoning that if an individual does not believe a group to be effective in solving a disadvantage—collective efficacy—or does not believe they as individuals are able to make a difference—individual efficacy—then there is rationally no reason to join a group or action (van Zomeren et al., 2008). Collective efficacy can be seen included in all the above-presented models and can interact with social identity in different ways.

Its main interactions are in group formation and its development from a group (Fritsche et al., 2018; van Zomeren et al., 2008). In other words, with a belief of collective efficacy, individuals can decide to join a group, or identification with a group can create or strengthen a belief of collective efficacy. Regardless, this perceived collective efficacy is shown to have a positive effect on collective action and is represented as such in the three models covered. Outside of these frameworks, similar findings indicate the increased inaction with a perceived lack of efficacy (Xiang et al., 2019) and the possibility to increase action intention through messages of collective efficacy (Jugert et al., 2016). This messaging was shown to increase individual and collective efficacy perceptions specifically in environmental realms. By communicating the efficacy of acting in a group in the face of climate change, collective efficacy beliefs also transferred to individual efficacy beliefs and increased their action intention. Previous successes in action, as well as observing others experience successes in similar actions may increase these efficacy beliefs on an individual, as well as a collective level (Fritsche & Masson, 2021). Individual efficacy as such, can play a role in interaction with the group and originating from it. Furthermore, it can play a relevant role through participatory efficacy—the belief that one's participation can benefit efficacy. This participatory efficacy has been found to be a stronger predictor of participation intention than solely collective efficacy beliefs (Bamberg et al., 2015). So, while collective efficacy is the interacting factor with social identity in collective action, participative efficacy might determine participation intentions more strongly. If an individual does not believe their participation will have an impact (in collective efficacy), there is less motivation to join.

3.8   Emotions

Reflecting on the SIMCA, EMSICA, and SIMPEA, a common thread that relates to how individuals and groups feel or perceive the world is what is most commonly called injustice perceptions. Moral convictions, as strong and absolute stances on moral issues (van Zomeren et al., 2012) similar to the moral obligations mentioned previously, fall in line in the same vein, where injustice would be perceived when something goes against such convictions. This can be both the way in which issues are brought forward and how collective action is animated (van Zomeren et al., 2012). Moral convictions are inherently linked to norms. When an issue is perceived as a threat to those norms that make up the social identity, individuals and groups can feel under attack and thus be persuaded to act. While environmental threats or threats to a social identity do not fall directly into this same category, they are the defining factors in the SIMPEA that play virtually the same role in determining a worthy cause to fight (Fritsche et al., 2018; Haugestad et al., 2021). Previous research suggests that recognizing environmental threats holistically, in their multiplicity of scales and interaction with social justice issues, can even be more effective in encouraging committed action than looking at them in isolation (Fisher, 2016). This holistic view might be a way to bring issues closer to the individuals or communicate the scale of the issues to increase injustice perceptions. Injustice perceptions in their effect on collective action, similarly play a role in radicalization processes (Jansma et al., 2022). These dynamic processes can escalate and deescalate the type of action taken, as they play a large part in the way people feel and through that the scale at which they react.

The way people feel is intrinsically intertwined with their perception of injustice, which guides their assessment of what presents as an issue and what does not. Personal and collective emotions can amplify various predictors of collective action (Landmann & Rohmann, 2020). While anger can play a part in motivating environmental action in situations like the *Hambacher Forst* clearance for brown coal mining in Germany (Landmann & Rohmann, 2020), research has also found prominent evidence for the effect of collective guilt on motivating environmental action (Haugestad et al., 2021; Rees & Bamberg, 2014). Furthermore, a sense of agency as mentioned in connection with the PCI (Simon & Klandermans, 2001) can also play a role. Collective guilt can be a manifestation of shared responsibility felt for climate issues, which were found to be crucial aspects in participation decisions in youth climate protests (Haugestad et al., 2021). Guilt as covered here, is inherently tied to group norms and morality as well. Knowing that one might have caused issues or played into systems that cause problems that go against one's norms puts into question one's identity and might as such evoke feelings of guilt. While other emotions might motivate more broad collective action, guilt lends itself especially to the environmental movement, as most individuals are bound to have contributed to environmental issues (Rees & Bamberg, 2014). So, if individuals or groups take on norms that aim for a more just treatment of people and the earth (in contrast to worsening climate catastrophes), they have likely inadvertently, in their past or present actions gone against those norms and might be motivated by guilt to right this discrepancy through collective action.

Furthermore, Landmann and Rohmann (2020) found that a feeling of being positively moved by action can motivate collective action. This may be a way that feelings play into efficacy considerations as well. Efficacy perceptions are not objective measurements of how effectively a group can achieve goals, though some logical concrete measures might play into it. It is more so feelings like the ones suggested by this study, where a feeling of being moved from seeing collective action can validate its efficacy in a subjective, yet tangible way. So, emphasizing this possibility to achieve something, might be a way to motivate collective action, as it elicits these emotions (Landmann & Rohmann, 2020).

This theoretical background does not aim to be a full account of all existing research on the motivations involved in collective action participation as there is a plethora of literature that goes into depth on a variety of specific topics, models, and predictors. Rather, it should give an overview of some of the significant points that have been made in the recent past and give insights into the complexity of the issue. It aims to provide a theoretical base, without limiting the possibility of finding points of contradiction or expansion when it comes to environmental activist groups. The discussion will connect the results of the study to the relevant theory and test the applicability of the models.

Methodology

4

4.1 Research Aim

This study entails an investigation of various narratives on the decision-making process of joining XR. Through narrative interviews, it aims to explore how people experience different factors that influence them to (not) participate in collective action. As environmental activism groups present a rather unique social group due to their formation around issues that are both structural and incidental, it is especially pertinent to investigate the motivations to participate. Moreso for a group that engages in slightly more radical collective action—XR. As addressed in the previous section, this social group engaging in collective action shows some similarities to previous research on collective action, but it does not seem to fit entirely into the presented frameworks. Environmental activism is unique in the type of fight and group formation, which previous research does not investigate specifically or independently. As environmental issues are currently very relevant to investigate, the motivations of people to join a group like XR can provide vital insights into the recruitment of movements that are crucial to the functioning and furthering of society. Thus, this research aims to investigate the applicability of the existing literature and to find points of expansion and adjustment. It aims to answer the following questions: *What patterns can be identified in the decision-making process of people joining the environmental activist group Extinction Rebellion in Stuttgart, Germany? To what extent are the existent models SIMCA, EMSICA, and SIMPEA applicable to this environmental activist realm? How does the significance of social identity manifest itself in the decision-making process?*

Supplementary Information The online version contains supplementary material available at https://doi.org/10.1007/978-3-658-44047-3_4.

The more unique form of collective action that XR represents in its approach and environmental nature, calls for a holistic research design that rather than prescribing specific narratives, allows individuals to explain their story and experiences as they see fit through narrative interviews only influenced by prompts if necessary.

The models presented previously were not tailored to the environmental movements and should therefore not frame the research preemptively. Therefore, the initial interviews should at first give standalone insights into decision-making to join more radical environmental collective action. Only if prompts for storytelling are necessary, prompting questions are posed that take inspiration from previous research. The gathered insights will then be deductively coded, and possible patterns extracted. While some bias from previous literature will naturally prevail through the deductive coding, it allows for this research to build on and expand previous theories and models.

The interviewees were initially sought through an open XR meeting, which allows anyone to join a meeting, be introduced to XR, and jump-start participation. As participating in a meeting is already an active behavior rather than solely attitude or intention, it allowed for samples of active XR participants—newly joined or established members. Further interviewees were sought in the existing group, through a snowball sampling method or through the Instagram followers of the local XR group, as this allowed for a clear connection to XR and built upon some garnered trust. It also allowed for interview participants that would not currently attend an XR meeting—individuals that decided against joining despite climate interest and knowledge of XR. The open meeting provided an accessible and non-intrusive insight into the group. Permission to join the meeting as a researcher was asked beforehand and this role was made clear to everyone present. By considering contextual factors through observations and field notes, insights can also be gained into how the XR group goes about motivating people to join and how they approach such an open meeting. By looking only at the Stuttgart XR group, local differences between groups and the influence of such differences on motivation to join are eliminated.

Reflexivity or self-awareness in this research is ensured through a clear statement and awareness of possible biases, the field notes, which keep track of possible influences of biases, and conferences with peers on interpretations if necessary (Morrow, 2005). As the investigating researcher has an academic, as well as personal history with environmental issues and movements and is a young German researcher that supports the movement, there is a clear bias with which the research is approached. This support, though not stated towards the research subjects, can still come across indirectly and thus influence how they might act.

This could be beneficial to gain trust and more open insight into their experiences but could also reduce objectivity.

4.2   Narrative Interviews

Narrative interviews (Rosenthal, 2014) were used to allow the interviewees to independently present the most important influences in their decision. Following the independent retelling of their experiences, the narrative interview style allowed for follow-up questions or further narrative-initiating questions, which could build on previous theory and prompt further influences for the interviewee. The narrative interview is based on the assumption that a person's experiences are most substantially represented in retellings, as it allows the person to re-experience the relevant processes, bringing them closer to their thoughts and feeling from that time period. It prioritizes the interviewee's perceptions of relevance and interpretations initially without the influence of the interviewer (Rosenthal, 2014). Previously, this method was mainly used in biographical research. However, it is also applicable to the reproduction of certain processes of action and chains of events of a person in their life (Rosenthal, 2014). The narrative style through broader retelling can contextualize experiences in wider personal history or societal processes. Following the examples of previous studies in this field (Bührle & Kimmerle, 2021; Fisher, 2016), these more narrative-focused interviews are well suited to the exploration of such subjective psychological decision-making processes, making apparent what was important to the interviewees. Follow-up questions into certain aspects of the decision-making process after the initial narration then allowed for elaboration on aspects that might have been forgotten or considered less important, for more holistic interview results.

To specify, the narrative Interview is a form of interview in which the interviewee can answer an initial question or story-telling prompt without interruptions or guidelines (Rosenthal, 2014). It is well suited for the present study to explore lived experiences of how interviewees make the decision to join XR. This specific chain of events in their past is unique to the individual and only through their retellings is it possible to elucidate the true importance of various influences on their decision-making. This style of interview is structured in three phases (Rosenthal, 2014):

1. First, the interviewee is given the opportunity to retell their experiences based on the initial prompt and/or question without interruptions.

2. Secondly, only after a clear conclusion of this retelling, follow-up questions are brought up, based on notes kept by the researcher throughout the interview. In this phase the questions are only limited to what has already been brought up by the interviewee, either pertaining to clarifications or elaborations on certain aspects.
3. Thirdly, new topics that can be relevant to the research can be brought up by the interviewer.

This structure can be valuable in the analysis, as the independent narration allows interviewees to bring up topics as they see fit. At which points certain topics are brought up and what topics might not have been brought up can further reveal additional information (Rosenthal, 2014).

The present study followed this narrative interview structure, starting (after a short introduction) with the prompt "Could you tell me about your decision-making process for participating in XR, what played a role, and what experiences you had? What were the rationales for and against joining?" for participants in XR and "Could you tell me about what is stopping you from participating in XR? What are the deciding influences and how did you come to that decision? What were the rationales for and against joining?" for individuals that decided against joining XR. Once the interviewees concluded with their retellings, follow-up questions pertaining to the retelling were asked. Finally, questions that were based on the theoretical background and built on previous studies (e.g., Rees and Bamberg, 2014) or further contextual questions were asked if the topics they refer to had not already been brought up. One such question was "What do you believe XR can achieve? What do you believe you could achieve as a part of XR?". This question was created to prompt efficacy considerations. The first question in this pair investigates collective efficacy considerations such as the ones suggested in van Zomeren et al.'s SIMCA (2008) and Fritsche et al.'s SIMPEA (2018). The second question aims to elaborate on the participatory efficacy aspect, suggested by Bamberg et al. (2015) to overshadow other efficacy considerations. Further questions concerning norms and morals, perceptions of the group, social influences, and emotions were also included.[1]

Some limitations for narrative interviews should be considered. One point of critique is that the willingness to narrate might not be guaranteed. While some interviewees could give extensive stories about their experiences, others might not

---

[1] The question protocol including the additional follow-up questions based on the literature from the *Theoretical Framework* chapter can be found in Appendix III (both in German and English).

feel as comfortable narrating freely and only give short responses. They might also find it difficult to piece together a narrative on the spot. These will need more prompts for a productive interview, which practically negates the benefits of the narrative interview. The influence of the previous research will be much stronger, as those topics will be specifically addressed through questions early on. Another point that might prove to be an issue is that decision-making processes are likely often not viewed as a concrete event, which could limit an individual's ability to form it into a narrative. Furthermore, as individuals are likely to bring up different points independently of the theoretical background, these might not be comparable between all interviewees in the analysis.

The research included 7 interviews that lasted between 13 and 33 minutes, 18 minutes on average. They were all conducted in German, according to language preferences. They were audio recorded and written consent to use the anonymized material recorded during the interview was obtained before or immediately following the interview. The interviews were transcribed verbatim thereafter following the rules of a simple transcription system (Dresing & Pehl, 2015). These rules include some commas to indicate speech flow and indications of unintelligible or unclear parts through '(unv.)' (unintelligible) and a word in brackets with a question mark if it is unclear.[2]

4.3 Observations

Initiating the study at an open meeting of XR presented the opportunity of supplementing the research with observations as a study method, allowing for triangulation. The additional insights through observations can present an added depth by discerning processes, organization, relationships, and interactions within XR (Thierbach & Petschick, 2014). This contextual information could be useful in better understanding the environment in which some decisions about joining are made. The observations for the current study were done openly for ethical, as well as practical reasons. As the meeting was used to find interview partners, the identity and intentions were made transparent before and during the meeting. Though most of the observations were done unstructured and as a passive participant, as the meeting was small it was not possible to fully refrain from partaking. This participation was limited to introduction and conclusion rounds.

---

[2] The transcribed interviews can be found in Appendix IV–X in the electronic supplementary material.

Field notes during and after the interviews also served as a method for a variety of sources, contextual information, and reflexivity (Morrow, 2005). These notes included such factors as the place of the interview, possible distractions or influential factors (sound, sight, etc.), and the time of the interview.

The triangulation of methods—use of narrative interviews and observations—is used to ensure a full picture of the multifaceted dynamics of the group and context for the individuals' decision-making. Triangulation involves the expansion of the approach to the research, in this case through a second method. This increased diversity of results gathered is not meant to act solely as confirmation of results and rather is utilized to gain further contextual insights (Flick, 2014). Observations lent themselves to this primarily because joining the meeting was required to find interview partners. This presented an opportunity to combine this search with additional results. The interviews with individuals were able to provide more detailed and personal insights into decision-making processes, while the observations during the meeting and the interviews could add both to reflexivity and to the contextual information, as well as the structure and inner workings of the local group specifically. These two methods were chosen to work together, as they were minimally disruptive or invasive to the group in comparison to other methods, personal involvement could ensure an establishment of trust, and they could work together seamlessly.

4.4   Deductive Coding

A directed approach to content analysis was used to analyze the interview transcripts. Previous research informed some of the coding but did not limit the opportunity to find new categories and aspects that might contradict it. While remaining qualitative in the analysis, through its firm steps and verifiability, the qualitative content analysis adopts some advantages of quantitative analyses (Mayring & Fenzl, 2014). This approach is well suited to the research, as its strengths lie in both supporting and extending existing research (Hsieh & Shannon, 2005). So, while some theory in a similar direction already exists which might be partially applicable to the present research, the directed approach addresses the need to build upon and extend this area of research further. Basing the analysis on existing theory allows for the testing of the applicability of existent models, but also has clear limitations, as a certain bias to find similar patterns is unavoidable (Hsieh & Shannon, 2005). Certain questions might also be formulated in a way that point to a certain answer or connection that interviewees then might be compelled to follow, though this should be less present in the narrative style interviews. Hsieh and Shannon (2005), lastly emphasize the loss of contextual aspects that can be another limitation to a strong focus on the theory. To attempt to counter these limitations, this study will attempt to counter a bias for the theory by explicitly searching for statements disproving the connections suggested in the theory. The narrative style interview allows a retelling without influence at first, which should eliminate the influence of the theory on the answers. This is, however, not fully avoidable, as the follow-up question or examples given might still suggest certain influences and connections and might have to be asked early on if the narrations are kept short. Lastly, observations and field notes serve to prevent losing vital contextual insights.

The deductive coding approach for the present study, in short, entails three phases. Initially, a coding scheme (Table 5.2, p. 37) is created to define the categories from the theory. Next, all parts of the material that refer to decision-making are determined and coded. Lastly, after coding with the existent codes, new inductive categories are created—then also included in the coding scheme—for the material that does not fit into the previously established codes. Further coding rounds are required thereafter. The category system is the center of the analysis (Mayring & Fenzl, 2014). The theoretical background will aid in determining initial categories based on key concepts and factors and how these categories relate to each other (Hsieh & Shannon, 2005). A coding scheme (Hsieh & Shannon, 2005) or agenda (Mayring, 2000) with category definitions serves to clearly distinguish between categories and ensures consistency. This is developed deductively and expanded on inductively. After a first read-through, the transcribed material was searched for text relating to the decision-making processes of joining XR. As the focus of the research, clearly defining what needs to be coded allows for more straightforward deductive coding and this method makes it very apparent where new code is necessary after initial deductive coding (Hsieh & Shannon, 2005). The coding (conducted using MaxQDA software) is a flexible, continual process involving many re-reads and some adjusting of the categories when necessary (inductive categorization). To avoid the theoretical background influencing the findings through confirmation bias, one of these re-reads aims specifically to find discrepant findings or evidence to disprove relationships suggested by previous research (Morrow, 2005). Themes developed through the categories can later be used in the interpretation of the results (e.g., Fisher, 2016).

Results

5

The research into the XR group in Stuttgart brought about extensive results that will be presented in this chapter. Observations of the open XR meeting will be followed by the composition of the interviewees. Subsequently, the factors initially mentioned in the narration of the individuals will be covered to introduce the interviewees and be able to discuss the order in the following chapter. This is followed by the coding scheme, which will introduce the results from the coding analysis of the transcripts. It is structured by the codes presented through the coding scheme. The chapter will conclude with a consolidation of the results to provide an overview and lead into the discussion in the following chapter.

5.1   Open-Meeting Observations

On 02.03.2023 at 19:00 in Stuttgart-Vaihingen (a city district considerably outside of the center), one of the individuals involved in XR provided a space in a shared house with snacks and drinks for an open XR meeting, which was observed for the present research.[1] The space was a room in the basement, a very casual, informal setting with a table, chairs, and two couches. While the meeting was advertised as an open meeting and an opportunity to get to know each other, it was more 'business' than expected. While a check-in round allowed for short introductions, the meeting seemed to be mostly a regular group meeting, which

---

[1] Appendix I in the electronic supplementary material includes the original observations.

Supplementary Information The online version contains supplementary material available at https://doi.org/10.1007/978-3-658-44047-3_5.

Y. Plate, *Social Identity Motivators in Environmental Collective Action*, BestMasters, https://doi.org/10.1007/978-3-658-44047-3_5

discussed recent group events, plans for upcoming actions, logistics, and new ideas. Outside of the researcher, there were 10 people in attendance in person and 2 online, which one person commented to be a lot, indicating the local group is rather small. This could be expected in a city with less significance than Hamburg or Berlin and in an action group less prevalent than FFF. It appeared three of the people in attendance were there for the first time.

The non-hierarchical structure was immediately apparent, as roles for the meeting—moderator, minute taker—were chosen at the beginning of the meeting. These are supposed to change from meeting to meeting, while bigger roles in the group are changed monthly. Seniority seemed to give more weight to contributions in things like decision-making. Meeting roles seemed to have a history of falling on the same people—those who more frequently volunteered and felt more comfortable in them.

The check-in/introduction round asked everyone to say their name, pronouns, how they were, and how they came to XR. This already gave an initial insight into the attendees' main motivations for participating. Four main reasons were named. The most named were previous engagements in various activism leading people to XR, mentioned by four people, and their social network introducing them to XR, also mentioned by four people. Two people also explained one of their main reasons being that they agreed with the morals of XR, and one person mentioned their agreement with XR's methods. This small insight already reflects some social, identity, and moral factors influencing individuals.

The meeting in total was about two hours long. A large part was discussing actions like the climate march happening the next day and actions for the coming month, allowing new participants to get involved immediately. For the more hands-on actions involving civil disobedience or 'bending' the laws, they do have action trainings before participants can get involved, also addressed in the meeting. An update on a citizens' council in Stuttgart being implemented gave a small impression of something the group might have effectively influenced. Indeed, one of XR's demands is the creation of such citizens' councils to involve the community in climate and environmental decisions and give them tangible power in politics. While the councils created in Stuttgart are not legally binding, they are still a step in the direction of achieving XR's goals and might lend a sense of collective efficacy to the group. Some of XR's morals and beliefs also came across throughout the meeting, starting with the inclusion of pronouns in the introduction round. The discussion of an inclusion workshop that just took place country-wide and a small activity in smaller groups discussing dominance structures and how to work against them additionally made clear some of the values the group holds and how they were actively working towards bringing those into their actions.

A brainstorming of ideas for the group generally and specifically for the coming month showed efforts to improve the group logistically (for example through better communication), brought ideas for future actions and suggestions to improve the incorporation of new members (for example through a buddy system). The meeting ended with a check-out.

While the meeting did not cater specifically to new members, it did allow them to gain an insight into the inner workings of the group and allowed them to find ways to join actions or meetings. Simply the involvement in the planning of future actions could give new participants the information they need to know where they can get involved. This open meeting also made apparent to new participants, the structure and dynamic of the group, as well as some of the morals and norms that are important to them. As there is no specific way to be an official member, the decision-making process is also not entirely clear-cut, and levels of participation in various actions and meetings can differ from person to person.

5.2   Composition of Interviewees

Interviews were conducted with seven individuals, three of whom were active in XR. Of the seven interviewees, five were activists and the other two had participated in activism in the past. The three XR participants were recruited for the interviews through the open meeting and two FFF activists through the snowball method. The two interviewees who had been active in the past were recruited through Instagram. The interviews were conducted mostly online through video calls, with only one occurring in person and one with solely voice call. Fieldnotes kept for each of the interviews[2] marked observations, possible interruptions, and influencing factors. While the interview in person had more distracting factors, as it was held in a public space, the video calls could have felt less personal, the voice call even more so. Table 5.1 shows the interviewees' ages, genders, occupation and/or education, and activism, with the short hands starting with XR being active XR participants and the short hands starting with N being non-participants. The composition of interviewees, as seen in Table 5.1, is mostly young (white) people, which fits with the general pattern seen in most environmental activism (Rainsford & Saunders, 2021). Additionally, several people who do not subscribe to the gender binary were interviewed, which can be attributed to generally more progressive views seen in left-leaning politics.

---

[2] Full field notes can be found in Appendix II in the electronic supplementary material.

Table 5.1 Interviewee Composition

Interviewees	Age	Gender	Occupation/Education	Activism
XR1	36	Woman/ Other	Various educations, quit Masters, now registered as unable to work	Founder of XR Stuttgart group
XR2	23	Other	Studying and active in the environmental realm	Active in XR
XR3	17	Man	In school	Active in XR
N1	22	Man	Worked for 5 years, now studying	Active in FFF
N2	22	Man	Studying and active in electoral politics (Bündnis 90/Die Grünen)	Previously active in FFF (& some XR)
N3	21	Woman	Studying	Active in FFF
N4	21	Other	Studying	Previously active in FFF

5.3   Introduction of Interviewees and Initial Factors

To give an initial impression of the interviewees, their stories are introduced, and their initial narration is summarized. The narrative interviews allowed for the individuals who are interviewed to first name the reasons they feel are relevant without outside prompts. These varied from person to person. The founder of the local XR group in Stuttgart (XR1) when interviewed, named their appreciation of the aesthetic and agreement with the values of the movement, as a significant reason for forming the local group of XR specifically. She also named the perceived injustice happening as her reason to be motivated to act. Answering the narrative question, she first mentioned her agreement with XR's descriptive and injunctive norms and then mentioned the lack of individual efficacy and the agency gained in groups. These were the points that first came to mind when asking her to recall how she came to start the local XR group.[3] Before joining XR, XR2 tried joining several other groups, which did not stick. They sought self-development through engagement in activism and took this as their primary motivation. As such, this was the first reason mentioned for joining XR. In response to the initial narrative question, they also recount their social connections to people engaged with XR,

---

[3] Their narrative question was slightly adapted to "Könntest du mir über deinen Entscheidungsprozess bei XR mitzumachen und eine lokale Gruppe zu gründen erzählen, was dabei für dich eine Rolle gespielt hat und welche Erfahrungen du dabei gemacht hast? Was hat dafür und dagegen gesprochen?", changing the wording of participating in XR to forming the local group.

which provided them with a positive image of the group. They further name their identification with the group and their agreement with their morals and injunctive, as well as descriptive norms. The last XR participant interviewed, XR3, was a very young and new member of XR, who also brought some of his friends to join XR. When asked why he decided to join XR, he first mentioned his agreement with their descriptive norms and further named his identification with the group. The motivator to become active for him was mentioned as well, naming frustration and perceived injustice. Lastly, he mentioned his doubt of the group's efficacy, which does not stop his active participation.

The non-participants were first asked what was stopping them from participating in XR. N1 has been active in FFF for over a year. He has friends that are active in XR but sees the group as relatively intransparent regarding their structure and dynamic. This is the first point mentioned, followed by his identification with FFF and belief in their efficacy, which is keeping him engaged in that group rather than XR. N2 used to be very active in FFF and participated in a couple of XR actions, but ultimately chose to go the electoral political route, which left little capacity for activism. He first talks about how the perceived injustice and climate crisis threat politicized him in the first place and motivated him to act. This brought him to seek agency through FFF, which he mentions seeing as efficacious through group action and which he found to have injunctive norms in line with his own. He continues by explaining how he saw his participatory efficacy to be higher in politics, which brought him onto his current path. Another FFF participant, N3, named a perceived lack of efficacy as one of the main reasons for not participating in XR. She does not see a connection between XR and FFF and feels there is not enough communication from XR about their events. This lack of an image and perceived lack of efficacy is what she saw as relevant when first asked what is stopping her from joining. She then explains how she believes FFF to be much more present and therefore more efficacious. Social connections within FFF motivated her to give FFF a chance and join an open meeting, while XR lacked such social connections for her. She finishes her retelling by naming the responsibility she was able to take on in FFF as a reason for her feeling increased participatory efficacy. The last interviewee, N4, had been active mainly in FFF in the past before moving in the direction of other groups such as *Ende Gelände* and *Antikapitalistisches Klimatreffen,* especially in the role of a photographer. They viewed XR quite negatively through their perception of the group's norms and morals. This was reinforced through similar views that friends in their activist circles held. N4 starts their response with their group involvement in FFF, taking on responsibility there and developing their politicized identity through activism. They mention the importance of the role social connections played in both their staying with FFF and their further activist involvement. While those social connections also existed with people involved in XR, this did not counteract their negative image. They emphasize that they do not agree with what they perceive XR's morals and norms to be and just do not identify with the group.

5.4   Coding Scheme

A total of twenty codes in six categories were utilized to analyze the transcribed material. The codes, as shown in Table 5.2, were derived from the literature, and supplemented with additional factors where necessary. The main categories: social identity, norms & morals, efficacy, and emotions can be seen in all three models covered in Chapter 3, but with the number of aspects these terms encompass, the concretization through the more specific codes is necessary to understand the entirety of the present patterns. The categories self-identity and other were added to accommodate additional literature and codes that came up when going through the transcripts.

The 15 codes derived from the literature (in blue; Table 5.2) were applied in the first deductive coding analysis of the literature. The code *group identification*, which plays a central role in social identity, was developed from the SIA (Reicher et al., 2010) and serves to reveal how individuals identify with a group and in what ways they see themselves as part of the group. As this identification is crucial to a shared social identity, this code aims to collect results that play into the group identification and its significance. *Outside social influences* is a code that was created to take into account the complexity of various social circles individuals might be a part of. It is based on Rees and Bamberg's (2014) study on perceived participation norms. While the name contains the word 'norms', the background of this factor is the role of social connections, which give weight to the norms. Thus, the focus of this code is on the influence those social connections have on the individuals' decisions, positive or negative. While the code *moral beliefs and values* was derived from the expanded SIMCA (van Zomeren et al., 2018), the codes *descriptive norms* and *injunctive norms* were derived from the normative alignment model (Thomas et al., 2009). While similar, these three codes still each serve their own purpose. Moral beliefs and values can be seen as more general, while descriptive and injunctive norms are more concrete, all specified in the definitions in Table 5.2 (codes 2.1.–2.3.). Perceived collective efficacy plays a role in all three models (Fritsche et al., 2018; Thomas et al., 2009; van Zomeren et al., 2018), and thus was applied as the code *group efficacy*. As a counterpart, *individual/participatory efficacy* was included to incorporate research by Bamberg et al. (2015) and Fritsche and Masson (2021), and its relevance in participation intention. *Injustice/moral conviction* is a code that is extremely prevalent in the literature in various forms (e.g., van Zomeren et al., 2008). It stands for perceived injustices, moral convictions an individual might feel, or the perceived environmental threat that is deemed to require action. *Group emotions* is a code created to represent the emotions originating from or influenced by social identity, as in SIMCA (van Zomeren et al., 2018). The codes *anger/frustration* and *guilt* were applied due to their relevance to environmental issues, as identified by previous studies (e.g., Haugestad et al., 2021; Landmann & Rohmann, 2020). *Agency/powerlessness* was founded in the PCI model (Simon & Klandermans, 2001) in which membership can lead to a feeling of agency. Powerlessness, a lack of a feeling of agency, acts as a counterpart. This addition is not based on the literature. While hopelessness and grief, as addressed by Jugert et al. (2016) was not named as an influencing factor in individuals' decision to join collective action, it does play a role in the perception of environmental issues and could be an interacting factor. Thus, it was added as the code *hopelessness/grief*. The code *politicized identity* represents the various studies on a politicized identity or opinion-based groups (e.g., Simon & Klandermans, 2001; van Zomeren et al., 2018). This includes individuals' politicization, political awareness, and action. Lastly, *support of self-image* was derived from the literature, developed from Fielding et al.'s (2008) findings that supporting one's self-image can be a powerful motivator and Bamberg et al.'s (2015) statement on historical participation motivating future participation.

Table 5.2 Coding Scheme

Category	Code	Definition
1. Social Identity	1.1. Group identification	Identification as a part of a group, significance of group for the individual and sense of belonging
	1.2. Outside social influences	Family, friends and acquaintances or the social environment and its influences on the individuals' decisions regarding activism
	1.3. Social connection and significant relationships	Social connection found and built in the group, significance of relationships within the group or influence of connections between participants & non-participants
	1.4. Group dynamics	Influences of the dynamics within the group; influence of structure, power-relations, and environment
2. Norms & Morals	2.1. Moral beliefs & values	Agreement with moral beliefs and values regarding principles, interactions, and priorities
	2.2. Descriptive norms (action)	Views on the groups' enacting of norms mainly pertaining to types of activism, structure of group, and strategy
	2.3. Injunctive norms (goals)	Views on a groups' big picture ideas, views, and visions for society, as well as critical viewpoints
3. Efficacy	3.1. Group efficacy	Beliefs in the group's efficacy to enact change, achieve its goals or be present in the media
	3.2. Individual/ participatory efficacy	Individuals' beliefs of their own ability to enact change individually or as a part of the group
4. Emotions	4.1. Injustice/Moral conviction	Actions motivated by a feeling of injustice or by feeling it is the morally right thing to do; a way to identify an issue that needs collective action to be addressed

(continued)

Table 5.2 (continued)

Category	Code	Definition
	4.2. Group emotions	Emotions shared, amplified, or mediated by the group
	4.3. Anger/Frustration	Actions motivated by anger or frustration
	4.4. Guilt	Actions motivated by guilt
	4.5. Agency/ Powerlessness	Actions motivated by seeking agency due to a feeling of powerlessness or hopelessness, as well as agency felt through participation in a group
	4.6. Hopelessness/ Grief	Actions influenced by hopelessness or grief
5. Self-identity	5.1. Politicized identity	An individuals self-identification in a political sense, including active participation in activism or in political realms trying to enact change
	5.2. Support of self-image	An individuals' actions or participation in groups to support their self-image as, e.g., an environmentalist
	5.3. Identity development	Individual motivation to develop their identity through participation in activism
6. Other	6.1. Responsibility	Roles taking on responsibility in a group can take and what effects this can have
	6.2. Image	The way a group is perceived or thought to be perceived and how this influences individual views and motivations to participate

The codes *social connection and significant relationships*, *group dynamics*, *identity development*, *responsibility*, and *image* were added to be able to cate-gorize sections of the transcripts that did not fit into the previously established codes or seemed necessary to address discussed topics (codes 1.3, 1.4, 5.3, 6.1, & 6.2; Table 5.2). Social connections and significant relationships, as such were not explicitly discussed in the literature but seemed significant for individuals' social identification. Group dynamics were mentioned quite concretely and thus required a separate category. It does, however, overlap with social identification and norms and morals. A group dynamic that fits into the individuals' moral perception could positively influence their identification with the group. Identity development was a completely new factor that was brought up during the interviews and did not fit into previous codes. Responsibility similarly emerged. It could increase group identification and participative efficacy and is thus a relevant additional factor. Image proved to be an important consideration as well and while overlapping with efficacy, also brings in social and moral factors. As such, it was also necessary to view this as a separate code. The following sections will present what these codes revealed in the research and what trends and reflections the interviews brought about.

5.5   Social Identity

In line with previous collective action models, social identity was an extremely present topic in the interviews. This category provided the largest number of codes, mentioned in some form by all interviewees. Group identification, outside social influences, social connections, and group dynamics covered a variety of social aspects involved in the decision-making process of joining environmental activist groups. As activism is a collective practice and humans are inherently social, these aspects naturally play a prevalent role for individuals. This role ranges from influences of established relationships to the identification with a group and their social dynamic.

Group identification
Group identification and the feeling of belonging in a group was only voiced distinctly a few times but can still be discerned from retellings and could factor in as a covert influence. Several individuals simply asserted their joining of the group and explained how they were active without explicitly mentioning their identification with the group (XR3, N2, N3, N4). When individuals already saw themselves as part of a different activist group, this often acted as a reason for not joining XR. N1, for example, explains:

> "[...] ich glaube das Problem ist vor allem, wenn man erst so spät von den Gruppen erfährt, dann hat man ja irgendwie schon seine Gruppe, wo man was macht, und steckt da vielleicht auch seine Zeit rein und kommt dann eher selten dazu bei was anderes noch was zu machen" (Appendix VII, p. 104, 6–9).

He sees it as a factor of already having *your* group and therefore not explicitly searching for another group and not necessarily possessing the capacities to get involved in an additional group. The identification with FFF, as *his* group, seems to also remove the incentive to search for identification within another group. He commented on FFF's *'chill'* atmosphere and having many people his own age, which seemed to contribute to him feeling comfortable in this group. While it is not always made explicit, individuals do talk about the group and their part in it in a way that conveys identification.

Nevertheless, some individuals did name this identification with a group or lack thereof very explicitly and as a very relevant factor in their decision. N4, in reference to XR, said:

"Das war einfach nicht meine Gruppe" (Appendix X, p.119, 11).

'It was just not my group'. They lacked identification with XR and recounted similar experiences of other friends and acquaintances from FFF, who got involved in XR but quickly realized it was not their group and moved on to different things. XR2 contrarily found XR to be exactly their group. After exploring different groups, where they felt the urge to retreat again, they found identification with XR. They explain:

"Und (…) insofern war es für mich glaube ich gar nicht so wichtig, welchen inhaltlichen Fokus die Gruppe hat, sondern eher zum einen, ob da Raum ist für solche Themen und zum andern ja, welche Menschen da so sind, wie da das Klima einfach so ist. Wie wohl ich mich da fühle" (XR2, Appendix V, p. 95, 2–5).

Prioritizing feeling comfortable in the group above the group's thematic focus shows that identification with the group took precedence in their decision. XR2 also identifies that taking on responsibility in this group helped them feel more like part of the group.

Outside influences
The activist groups (XR and FFF in this case) cannot be explored in isolation, as the individuals that participate are intertwined in various social networks with different views and expectations. Parents were the most named group for outside influences. They often had a more critical view of activism but did not stop the interviewed people from participating. XR3 responded to the prompt on social influences in his environment:

"Also, meine Eltern natürlich. Die waren überrascht. Also, dass es auch natürlich direkt, wenn man so das online eingibt, ist da natürlich auch immer direkt das Extreme sehr stark vertreten. Sie meinten, ob ich nicht vielleicht mal leichter einsteigen will" (Appendix VI, p. 100, 32–34).

Parents, when involved in their children's lives, are bound to be apprehensive of more radical groups. XR3's parents asking him if he would not want to start *'a bit lighter'* shows exactly this involvement and the tendency for parents to dampen the level of action their children might engage in. XR2 also considers:

"Aber ich glaube schon, dass meine Familie, zum Beispiel eher was dagegen gehabt hätte, wenn ich mich jetzt der Letzten Generation angeschlossen hätte. Ich weiß aber nicht, ob mich das jetzt davon abgehalten hätte, unbedingt. Aber ich glaube eigentlich schon, dass ich eher bestrebt bin Konflikte zu vermeiden, auch innerhalb der Familie" (Appendix V, p. 97, 34–38).

They convey the possible underlying factor of family expectations. Unconsciously, they might join a less radical group so as to not foster too much conflict within the family. The perceptions of the familial social circle are likely quite relevant to many individuals. This might decrease with more time spent away from home or vary per individual, as N1 relays:

"So gerade meine Familie hätte mich auf jeden Fall nicht auf eine Fridays for Future Demo gelassen. Die sind politisch sehr anders eingestellt" (Appendix VII, p. 104, 30–32).

His family would not have supported him joining FFF but after studying and educating himself on the topic, he decided to join anyways. This did not seem to largely affect his relationship to his family though. It is also possible that this upbringing already made FFF seem quite radical, furthering the previous point on dampening participation in activism. Other parents might be more inclined to support or understand their children's actions. XR1 recounts various influences when first deciding to start the local XR group:

"Es gab schon Menschen, also keine jetzt direkt Bekannten oder Freunde von mir, […] die sich sehr kritisch geäußert haben. (…) Es gab aber schon mehr Menschen aus meinem Umfeld, die verstehen konnten, dass ich mich da so reinhänge, weil also meine Eltern sind im Prinzip auch so, oder waren so, als sie jünger waren, nur halt mit einem ganz anderen Thema, aber die waren auch so, dass sie sich ja einer Sache verschrieben haben und da wirklich, könnte schon fast sagen monothematisch dann unterwegs waren und auch viel bereit waren dafür aufzugeben oder zu opfern. Also aus der Richtung gab eher, ich weiß nicht, ob sie das direkt verstehen konnten, aber nachvollziehen auf jeden Fall, warum" (Appendix IV, p. 91, 32–40).

While she did experience some negative pushback, the people close to her were much more understanding. Her parents especially, as they had had similar motivations in their youth and could therefore relate to her passion. Generally, parents were less likely to dissuade action, but might inadvertently influence their children to join less radical groups.

Friends outside of the activism circles seemed to have an influence less often. XR1 recounts:

"Aber so aus dem Freundes- und Bekanntenkreis, den ich damals hatte, sind tatsäch-lich nicht so viel übriggeblieben, weil sich einfach die (...), ja, die gemeinsamen Themen quasi aufgelöst haben" (Appendix IV, p.91, 40–42).

While they seemed rather understanding when first starting the group, this outside influence dissipated, as friends likely formed increasingly inside activist circles, where they shared more topics and values. N4's influence by friends is partially intertwined with their social connections. Their fellow activists in FFF influenced them against XR by sharing their negative image. They explain:

"Ja. Also, in so den Kreisen mit den engsten Genoss*innen, mit denen ich darüber geredet habe oder wo das Thema war, war es schon eher so, dieses sich über XR lustig machen und sagen, das sind ja die, weiß nicht, die die Esofritzen aus der Klimabewegung so bisschen" (N4, Appendix X, p. 121, 16–19).

This general attitude of making fun of XR seemed to be a strong influence on their own negative image of the group.

The last outside social influence, named twice, is the individuals' schools. N2 and N3 both recounted how they were politicized through climate topics through their schooling and how this led them to become politically active. N3 remembers:

"Das hat in der Schule schon angefangen. Also bei mir an der Schule gab es so ein breites Feld, wo man sich sozial engagieren konnte. [...] Es war auch Pflicht für jede Klasse, mindestens einmal pro Halbjahr auf einer Fridays for Future Demo zu sein, was ziemlich cool war natürlich. Und was dann schon so ein Interesse geweckt hat" (Appendix IX, p. 115, 9–14).

She was already confronted with many opportunities in school to become socially engaged. This amount of opportunity and openness at school allows students to get involved much more easily and possibly removes barriers. By making it a requirement to join a FFF demonstration twice a year, it likely makes it much less intimidating to participate in more activism.

Social connection and significant relationships
One of the most significant factors—and an inductive code—in the social identity category was the social connections within the group and the significant relationships to people in a group. These connections were often a determining factor for joining activist groups or for staying with a group. N3 recounts:

> "Darüber hinaus durch die ganzen anderen Aktivitäten, die ich habe, also in anderen Bereichen Aktivismus betreibe, kannte ich ja auch schon Leute, die bei Fridays for Future sind, was natürlich schon auch ein schon größerer Faktor war. Hey, die Leute sind cool, ich kenne die auch schon, also schnuppere ich mal kurz rein, um zu schauen, wie es so läuft [...]" (Appendix IX, p. 113, 31–34).

Her connections to people in FFF brought her to give the group a try, while such connections to XR did not exist for her. She says:

> "Und ja, wie gesagt, also bei XR kenne ich leider keinen Menschen, habe da auch nie reingeschaut oder so" (N3, Appendix IX, p.113, 38–39).

Quite clearly, this was a large factor for her to explore FFF and similar connections to XR would have likely had a similar effect. Already knowing people in a group might provide advantages like inside knowledge and a more comfortable introduction to the group. In the open XR meeting, four people also mentioned their social network bringing them to XR. XR3 specifically also mentions:

> "Also und tatsächlich habe ich dann einige eingeladen, die jetzt auch bei Extinction Rebellion sind, aus dem anderen Umfeld" (Appendix VI, p. 100, 38–39).

His invitation as an XR member convinced his friends to also join the group. Some of those friends were also present at the open meeting, joining for the first time. Furthermore, XR2 explains:

> "[...] und dann hat auf jeden Fall eine wichtige Rolle gespielt, dass in meiner WG halt Menschen waren, mit denen ich mich gut verstanden habe, die auch schon in verschiedenen Gruppen aktiv waren. Unter anderem auch XR. Und ich von daher schon

immer wieder halt ein bisschen was erzählt bekommen habe und überhaupt quasi von
XR erfahren habe und dadurch auch ein gewisses Bild von XR bekommen habe"
(Appendix V, p. 94, 32–36).

They were living with people who were involved in XR, which allowed them to
learn about it and garner a positive image. This relationship with people inside
the group ultimately acted as positive encouragement to explore the group further.
This is not guaranteed though. N4, who was active in FFF for a while, did not
get involved in XR despite having connections to people in the group. They
only found out about the group through friends and fellow activists after getting
involved in activist circles but did not gain a positive image through this, as will
be explored further in the section for the corresponding code. They do however
narrate:

"Und ich glaube auch zu den anderen Gruppen, zu denen ich dann bin oder wo ich
dann auch mitgemacht habe. Das war dann auch nur über die Leute, die ich kannte und
mit denen ich dann gerne was zusammen gemacht habe und so was" (N4, Appendix
X, p. 121, 8–10).

This shows that relationships with people in other groups did convince them
to explore those groups more, while similar connections did not have the same
effect for XR. The negative image might have had more weight here than social
connections. N1 also shows that his connections to XR are not enough to convince
him to participate. He relates:

"Tatsächlich finde ich, dass bei XR vieles ein bisschen intransparent nach außen
scheint. Also ich als Außenstehender, obwohl ich viele Freunde auch bei XR habe,
weiß da tatsächlich sehr wenig drüber" (N1, Appendix VII, p. 103, 28–30).

His connection to the group alone is not defining. N2 similarly had strong connections to various activists that advocated for their path, but parallel to that he
had friends engaged in politics that were advocating for their path. He cannot say
for sure how they influenced his decision, but his connections to activism ultimately did not convince him to stay there. N2 saw his relationships with people
in activism quickly wither when he chose the electoral political path for himself,
as they did not agree with his decision. This would also make it much harder for
him to join actions now. In this case, the lack or loss of relationships is a factor
that can hold him back from participating in future actions, further supporting
that social connections play a role in participation. N1, N2, and N4 also describe
the often-present interaction between FFF and XR, with N2 also joining some

XR actions during his FFF-past. N3 claims the contrary. She sees no connection between the two groups. With these differing perspectives on relationships between people in FFF and XR, it seems to be very dependent on the individual and their connections. The extent and form of the relationship an individual has with someone in an activist group likely also determines its influence on their decision-making. So, while relationships with people that are a part of activist groups can be a determining factor for people to explore those groups further, they are not predictive on their own.

Social connections within a group are of similar importance, though usually not as defining in the initial decision-making process. These might have a stronger role in retaining involvement in groups. N4, who did not know anyone when they first joined FFF, still emphasizes the social connections they made within the group and how they led him to stay with the group longer than they would have otherwise. They recount:

> "Ich hatte gegen Ende eigentlich auch keinen Bock mehr, keine Energie mehr irgendwas für FFF zu machen, aber weil da halt meine engsten Freund*innen waren, habe ich halt trotzdem weitergemacht" (N4, Appendix X, pp. 120–121, 45–3).

And:

> "Wenn man jeden Freitag eine Demo organisiert und halt richtig viel zusammen durchmacht, dann schließt das auch total zusammen und man hängt auch total an den Leuten" (N4, Appendix X, p. 121, 6–8).

The strong social connections formed within a group and how those might sway individuals to stay with a group are apparent in these statements. The shared experiences mentioned can factor into building up a social identity. This can even be done quite deliberately. XR2 narrates:

> "Ja, ich glaube die erste Veranstaltung, wo ich bei XR war, war tatsächlich auch ein Resilienztreffen, wo wir uns draußen im Park getroffen haben und halt auch sehr persönlich und sehr schön uns irgendwie ausgetauscht haben und Kraft gegeben haben und ich denke schon, dass mich das auch positiv beeinflusst hat" (Appendix V, p. 95, 24–28).

XR organizes specific events for their members to bond and share experiences and emotions. In the case of XR2, forming these social connections was a positive influence in their decision to join the group. This was likely only strengthened by the fact they were specifically searching for more social connections after feeling lonely due to the COVID-19 pandemic. The social connection is a priority to them, as they also recite:

> "Ja, ich denke es ist schon auch wichtig irgendwie ja dann die Menschen zu kennen die da sind und zu denen irgendwie eine Verbindung zu haben. Und die einem wichtig sind" (XR2, Appendix V, p. 96, 15–17).

They reiterate how the social aspects are an important factor, which they value highly. XR3 agrees with this, as he also values the connections present in the group and praises the closeness within the group. The quick social connection positively influencing people to join can also be seen in N1's retelling about him joining FFF:

> "Wenn man aber einfach mal da ist und so danach dableibt und fragt so ‚Jo, hey, kann ich irgendwie was beim Abbau helfen?‘ oder so, dann sind das natürlich auch alles ganz normale Leute, die einen dann auch noch mitnehmen zum Essen oder sowas. Und daraus entwickeln sich relativ schnell Freundschaften, die dann eben beständig bleiben, so" (Appendix VII, p. 104, 32–36).

Connecting with FFF members after a protest swayed him to get more involved in the group. And the strong social connections he formed play a vital role for him, as he said:

> „Ja, also die einzigen Gründe warum ich wirklich aktiv bei Fridays for Future bin, sind halt der Druck, den man damit auswirken kann und dass es meine Freunde sind" (N1, Appendix VII, p. 104, 36–37).

Social connection, again, is a large factor for people to stay in their group. N1 emphasizes his friends in the group are one of the two main reasons he is participating in FFF. Social connections and relationships with people in activist groups are not a requirement for joining, nor are they sole predictors of decisions to participate. They do, however, seem to have the capacity to strongly influence individuals to join certain groups.

Group dynamics

Group dynamics were mentioned several times during the interviews, influencing how individuals felt about different activist groups. These dynamics could play into how much they identified with the group and how comfortable they felt in a group. Furthermore, they could play into norm and value factors where the way people act within the group should align with an individual's values. The dynamics in the local XR group were mostly praised by participants, with XR2 naming it as a reason to join and them looking out for specifically a good group *'climate'* and finding this in XR and XR3 saying:

> "Achso ja, ja die Gruppendynamik ist schon relativ eng auch. Es gibt auch teilweise Aktionen, die nur dafür da sind, eben diese Verbindung untereinander zu stärken" (Appendix VI, p. 101, 35–36).

The relaxed atmosphere at the open XR meeting supported this positive image as well. However, when discussing other meetings, especially meetings outside of the local group, XR2 considers:

> "Also ich merke, wenn bei einzelnen Treffen irgendwie deutlich vor allem männlich gelesene Menschen da sind, dann finde ich es irgendwie anstrengender. Und ich kann mir vorstellen, dass sich insbesondere FLINTA Personen dann nicht so wohlfühlen. Und ja, ich glaube auch, dass es von Ortsgruppen zu Ortsgruppen relativ unterschiedlich sind, was für Menschen da so sind und ich habe jetzt auch bei so Treffen wo aus vielen Ortsgruppen und Aktionsgruppen irgendwie Menschen zusammenkommen, also Austauschtreffen, festgestellt, dass da so ein paar ältere weiße cis Männer gibt, die für die Atmosphäre dann manchmal ein bisschen unangenehm sein könnten. Und ich glaube sowas kann quasi ein Punkt sein, der dafürspricht oder dagegenspricht. Und ich glaube ich selber versuche auch schon also, wenn es irgendwie Gruppen sind, die von männlich gelesenen Menschen irgendwie dominiert werden, dann habe ich da auch kein so großes Interesse da dazuzugehören" (Appendix V, p. 98, 13–24)

The atmosphere and dynamic can differ greatly depending on the local group and the people participating. XR2 here also reflects on how a male-dominated space likely negatively affects FLINTA (women, lesbians, intersex, non-binary, trans- and agender) people. When male-presenting people form the majority of a meeting, they perceive this as more strenuous and feel less inclined to participate. Additionally, older, white, cis men can negatively influence the dynamic. XR2 also explains how they feel the structure of the group influences its dynamics. The non-hierarchical organization can be more exhausting but also more rewarding, as every member becomes more involved and carries some of the responsibility. Although nobody is forced to join actions, people might still feel pressured to join more than they are comfortable with or take on more than they can handle. This might lead them to distance themselves from the group. Conflicts that can come up due to knowledge hierarchies or unclear responsibilities could create an uncomfortable environment for people who are sensitive to such things. And lastly, social hierarchies might be reflected in the group structure unwillingly, like the dominance of male-presenting people in some meetings. Overall, XR2 brings into consideration many aspects that can influence group dynamics and how those might affect individuals differently. It is a complex factor that was also addressed by FFF participants. N1 commented on the *'chill'* atmosphere in FFF but felt like he does not know what XR's group dynamic looks like. N4 comments:

> "Also es fühlt sich viel mehr von unten an als XR, die so einen Gründer*innenkern hat und der auch so persönlich in Erscheinung tritt" (Appendix X, p. 122, 22–24).

Naming the FFF dynamics, which feel more bottom-up to them as a reason they felt more connected to this group. Group dynamics can influence many aspects of individuals' experiences within groups and can affect how willing they are to actively get involved or stay involved.

5.6   Norms and Morals

Collective moral beliefs and values, descriptive norms, and injunctive norms were also very present topics in the interviews, with this category presenting the second most coded excerpts. Groups' norms and morals that coincided with the individuals' were a prevalent consideration.

Collective moral beliefs and values
Shared moral beliefs and values were not only named as a reason to participate in or avoid a group but were also named as a social aspect that was considered to be valuable. XR1 succinctly says:

> "Man braucht immer andere, die da auch mitziehen und die gleichen Werte teilen" (Appendix IV, p. 89, 42–43).

The shared values act as a prerequisite for a group to form. XR1 specifically names the focus on dominance structures and on handling emotions as valuable in XR. Sharing the values placed on reflection with XR was also a factor for XR2 to be swayed toward the group. XR3 adds that sharing values is not just necessary but also a benefit of the group:

> "Also es ist natürlich schön Leute mit den gleichen Sichtweisen zu treffen (unv.) und dann tatsächlich auch ja sich auszutauschen [...]" (Appendix VI, p. 101, 30–31).

Meeting people with the same beliefs and being able to engage with them was a very positive experience for him. These values can also be a point of how they are perceived, as it is up to individuals to interpret some of the specifics. N4 views the values contrary to most of the other interviewees, criticizing:

> "Es auch noch mehr so dieses, dass es irgendwie auch viel so Antisemitismusvorwürfe gibt gegen einzelne Personen oder viele Leute aus der Bewegung oft so ein unreflektierter Umgang mit irgendwie Rassismus und solchen Sachen sind halt oft" (Appendix X, p. 122, 7–9).

And commenting on XR stereotypes:

> "Es mag vielleicht alles der Stereotyp sein, aber es ist schon so ein bisschen die Sachen, die mich davon abgehalten haben, sind so dieses, dass es sich sehr oft sehr nah an was sehr esoterisch anfühlt" (Appendix X, pp. 118–119, 43–2).

This negative perception of XR values was a large factor in stopping N4 from getting involved in the group. Their outside perspective might also be shaped by actions from XR members going against the group's values (as stated in their principles; Extinction Rebellion, n.d.) or faulty execution of those values, pointing to more of a discrepancy in descriptive norms. It could also be attributed to a simple difference in values.

Descriptive norms
As the norms shown through action in activist groups, descriptive norms played a very relevant role in the decision-making processes. These present and visual norms play a role not only in what kind of action people are involved in the group but also in the perception of the group by the public. As XR's actions are largely based on civil disobedience, this factor is also a point where opinions can diverge. The XR members that were interviewed all mentioned that they agreed with XR's actions. XR1 states:

> "Ja, also mir hat das einfach sehr gefallen, der Ansatz gefällt mir sehr, ich finde, dass wir den unterwegs ein bisschen verloren haben aus verschiedenen Gründen, aber im Grunde genommen, so die Grundidee, finde ich immer noch super" (Appendix IV, p. 90, 21–23).

Naming the approach as one of the factors that spoke to her. This agreement with descriptive norms can also be seen in several of XR3's statements. He retells how he first found out about XR:

"Habe dann irgendwann einen Artikel gelesen. Ich glaube vom Greenpeace Magazin oder so, wo Extinction Rebellion dann tatsächlich erwähnt wurde. Und habe mich dann online darüber erkundigt was die so machen und dachte ‚Ja, finde ich cool'" (XR3, Appendix VI, p. 99, 26–29).

The actions that he read about were ones he found *cool,* which persuaded him to look into the group more. He further specifies:

"Und ich finde Extinction Rebellion jetzt speziell ist da ein gutes Mittelmaß zwischen beidem, wo auch, was ich schön finde so an Aktionen oft auch ein bisschen Kreativität eine Rolle spielt, also mit irgendwelchen Performances oder so" (XR3, Appendix VI, p. 100, 5–7).

He names XR as a middle ground between FFF and *Letzte Generation* and adds:

"Und so kann man sich dann praktisch die Aktion raussuchen, die man vertreten kann oder die zu einem passen [...]" (XR3, Appendix VI, p. 100, 23–24).

After finding out more about XR, he found not only the first impression to be something he would like to join but also appreciated the active engagement with creativity and the flexibility with which actions one wants to join. So, even if he would not agree with a specific action, he could just refrain from joining that one action. He does, however, reflect that the more radical actions that XR did were a point of hesitation for him to join. Contrarily, one person in the open meeting named these more radical actions as one of the main reasons for joining the group. XR2 names another aspect of XR's descriptive norms:

" [...] wir leben auch selber das Stück weit Utopie" (Appendix V, p. 96, 29–30).

They refer less to the form of activism and more to bringing together the descriptive and injunctive norms, where the ideals and goals of the group are already lived within the group.

XR non-participants generally did not have anything negative to say about the group's actions. N2, for example, was very open in his support, stating:

"Ich hatte bei Fridays for Future viel mit Extinction Rebellion zu tun. Und ich befürworte auch die Aktionen von Extinction Rebellion, also vor allem auch den zivilen Ungehorsam. Ich glaube, dass es notwendig ist in einer demokratischen Gesellschaft, dass es diese Aktionsform gibt und vor allem, dass sich auch so eine Gruppe für diese Thematik einsetzt und auch auf diese Weise für eine Thematik einsetzt" (Appendix VIII, p. 110, 32–36).

He makes it very apparent that their actions are not the limiting factor for him. He supports XR's actions, including the civil disobedience, and sees it as necessary for a working democracy. Nevertheless, the actions do limit him, as he explains that joining in more radical actions that go against the law or could result in arrest would risk his political future, as he is required to pass background checks. A similar pattern can be seen in N3's statement:

> "Und ja, ich persönlich würde sagen, dass ich keine Angst davor habe, also und schon gerne vielleicht bei Aktionen dabei sein möchte. Aber dafür muss ich halt von den Aktionen mitbekommen dann" (Appendix IX, p. 115, 36–38).

She agrees with the descriptive norms, and they are not the limiting factor. She reiterates that she does not find out about the actions, referring to a previous point where she conveyed her view of XR's lack of perceived efficacy. N4 on the other hand, recounts:

> "Aber ab dem Punkt, wo mir dann FFF nicht genug war, war ich halt auch an anderen Stellen meiner politischen Entwicklung an dem Punkt, wo mir auch XR nicht mehr genug war. Und dann halt auch wieder zusammenhängend mit so Umgangsweise mit Polizei und so weiter, mit der ich einfach nicht übereinstimme" (Appendix X, p. 119, 40–43).

N4 did not see eye to eye with XR. Furthermore, when they started feeling like FFF was not enough, XR fell in the same category. Additionally, they did not agree with their actions in areas like police interaction.

When referring to their own group, FFF participants also reiterated their agreement with their group's actions. N1 experienced this after going to his first demonstration, recounting:

> "Ich habe mir dann irgendeinen Kollegen geschnappt, bin dann dahin gegangen, das war hier in Stuttgart der fast größte Streik mit irgendwie 15 000 Leuten. Und ich fand das einfach richtig geil, dass die das alles organisiert haben, mit Bühne und so und viel Musik und diese (shouts?) rufen und dann bleibst du eben einfach dabei" (Appendix VII, p. 104, 19–22).

After joining a significant demonstration and liking this FFF action, it sparked his motivation to join the group more permanently. He adds:

> "Ich würde nicht sagen, dass es die beste Aktivismusform ist, weil der stille Protest auch nur davon leben kann, dass es einen nicht stillen Protest gibt" (N1, Appendix VII, p. 105, 6–7).

He clarifies that he does not necessarily view this as a better form of activism than others but does feel it is one of the best. This makes clear that he still agrees with XR's actions, their descriptive norms, but this is not motivation enough to join them. Overall, descriptive norms only proved a point of hesitation for one person. It was a factor that could be defining but was not predictive of joining. The lack of opinions against XR's actions could be attributed to the fact that all interviewees were or had been activists.

Injunctive norms
The strategy, goals, and focuses of activist groups, representing their injunctive norms, were named several times as important reasons to participate in groups, but were not mentioned by everyone. These norms can also be subject to perception, as even clearly stated norms could be misinterpreted or miscommunicated. While injunctive norms were not named as an influencing factor by XR3, the other two XR members that were interviewed named them as significant factors. XR1 explains from her view:

> "Ja, mir gefällt die Ästhetik und die Tatsache, dass es eben das erklärte Ziel war eine gesellschaftspolitische Komponente in den europäischen Klimaaktivismus zu bringen. Also XR sieht sich ja schon als Teil der Klimagerechtigkeitsbewegung, aber nicht so sehr als Klimabewegung, sondern eher als gesellschaftspolitische Bewegungen" (Appendix IV, p. 89, 33–36).

One of the main goals of XR is emphasized (intersectional climate justice). She refers to the nature of the movement and the group's purpose, which convinced her to become a part of it. These goals and the main 10 principles of XR were her tools in communicating the group's purpose on social media in an attempt at assembling a local group. XR2 names the foci of the group as an important factor for deciding to join as well. They relate:

> "Ja, ich glaube mir persönlich ist wichtig, dass sehr viele Themen da eigentlich zusammenhängen, also dass es nicht nur um Klimagerechtigkeit geht, sondern, dass da Feminismus eine Rolle spielt, dass da Antirassismus, Antikolonialismus eine wichtige Rolle spielt, Inklusion, Awareness und so weiter" (XR2, Appendix V, pp. 94–95, 42–2).

They did not only find the topic of climate change important but specifically the connection between various societal issues and the focus on addressing them and their connection. The principle of XR that everyone is welcome was also named as a positive factor. This agreement with XR's injunctive norms seemed to play a vital role in their decision-making. Two people at the open XR meeting named this as a defining factor as well. Their agreement with XR's morals was one of their main reasons for joining the group. XR1 also claims:

> "Aber gerade das, was in den Prinzipien festgehalten ist, ist ja so das Gerüst oder der Kleber, der uns zusammenhält und ja die fand ich von Anfang an toll, weil sie schon, ohne das so explizit auszudrücken, darauf abzielen unsere Gesellschaft also das Miteinander komplett zu verändern" (Appendix IV, p. 91, 21–24).

She believes that this agreement—recorded in their principles—is what unifies the group. N2's statement shows once again that the agreement with the group's injunctive norms does not have to be determining in individuals' decisions:

> "Aber die andere Sache, inwieweit ich moralisch und vor allem auch thematisch dahinterstehe und das ist bei mir, bei Extinction Rebellion, auf jeden Fall der Fall" (Appendix VIII, p. 111, 16–18).

But even though he stands behind their norms and morals, he chose a different path for himself. This factor was not his priority in his decision-making. N4, contrarily, did see this as a priority, but the perception they had of the group spoke against their agreement. They saw the norms as a strong point against joining, recounting:

> "Wie mit der Klimakrise umgegangen wird und vor allem zu der Zeit, wo ich es mit-gekriegt habe, zumindest, war es ja doch oft so, dass zum Beispiel XR nur auf das Thema Klima fokussiert hat und nicht die multiplen Krisen drum herum irgendwie mit einbezogen hat in die Aktionen und Öffentlichkeitsarbeit, die gemacht wurde. Und es hat mich immer sehr davon abgehalten bei XR mitzumachen" (N4, Appendix X, p. 119, 5–9).

This might also speak to a discrepancy between injunctive norms and how these were portrayed as descriptive norms.

5.7   Efficacy

Group efficacy
Efficacy proved to be a surprising factor pertaining to XR, as it was often not a necessary factor to participate. While it did show to be deciding for mainly

people involved with FFF, it was often not a focus of people's decision-making process for XR. XR1 at first explains:

> "Also, Extinction Rebellion wurde ja in UK von einer ganzen Gruppe gegründet. 15 oder 20 Menschen, die haben mehrmals versucht eine Bewegung zu starten immer mit anderen Ansätzen, eine andere Kommunikation, andere Zielgruppen und so und Extinction Rebellion hat letztendlich funktioniert" (Appendix IV, pp. 90–91, 45–3).

XR was a group model that seemed to work and gain traction. She uses this history of the group, which was established out of several attempts to form working activist groups, to explain her belief in the group. Her phrasing of the group 'working' would imply some form of efficacy. When asked about efficacy though, she clarifies:

> "Ja, also diese Hoffnung auf eine wirkliche Wirksamkeit, wie wir uns das so allgemein vorstellen, habe ich ziemlich schnell aufgegeben. Mehr über die Erfahrungen mit Fridays for Future als jetzt die Erfahrungen mit der XR" (XR1, Appendix IV, p. 92, 24–26).

Seeing FFF fail despite their reach, momentum, and numbers, took away her hope for true efficacy. XR3 shares this lack of collective efficacy belief, he emphasizes:

> "Wenn ich ehrlich sein soll, bin ich mir nicht sicher, ob das überhaupt eine Wirkung hat, aber ich möchte es zumindest versucht haben" (Appendix VI, p. 99, 37–38).

He is not convinced of the group's efficacy, but he at least wants to give it a shot. With such a systematically engrained and wide-reaching problem as climate change, the actual solving of the issue is near impossible. Focusing on their own engagement might alleviate some of the perceived lack of group efficacy. Despite not being convinced of XR's efficacy, XR3 holds out hope that it could be going one step further. He considers:

> "Und bei dem Globalen, es ist halt so, dass man irgendwann das Argument von Letzter Generation oder so, dass Fridays for Future auch nichts gebracht hat und dass man jetzt dann halt (nächster?) weitergeht also ja, so denke ich halt, ob es mehr bringt oder nicht weiß ich nicht" (XR3, Appendix VI, p. 101, 8–11).

He thinks critically of FFF's efficacy and sees XR as the next step to try to affect change although he says he does not know if it is more effective or not. XR2 tried not to consider the efficacy of the group at all in their decision, stating:

"Also genau, ich habe glaube ich schon versucht da nicht so drauf zu achten. Also ich denke, dass es sowieso schwer zu sagen ist was tatsächlich eine Veränderung bewirkt, und oft sind solche Veränderungen halt auch nicht wirklich zu sehen" (Appendix V, p. 96, 39–41).

This factor was not a priority and they put more focus on doing their part. Altogether, XR members did not necessarily believe in the group's efficacy but placed more importance on acting on the perceived injustice and standing up for what they believe in. FFF participants placed much more of a focus on their perceived collective efficacy.

The perceived higher efficacy of FFF in comparison to XR came up more than once, as a reason to stay with FFF. N3 explains that she sees FFF as much more efficacious and contrastingly speaking on XR's efficacy:

"Und zweiteres wäre halt, dass ich leider, es kann ja auch sein, dass die richtig viel machen hier in Stuttgart, aber dass ich halt leider relativ wenig mitbekomme und, dass ich dann auch irgendwie da drin leider kein Potenzial, also ich persönlich, also hier in Stuttgart zumindest, kein Potenzial drin sehe" (Appendix IX, p. 114, 22–25).

She feels as though FFF has a much larger presence and since XR is much less visible to her, she does not see any potential in the group even though she is supportive of their actions. N1 has a similar view, where he does not feel XR to be very present and sees more efficacy in FFF. He illustrates:

"Ob das jetzt eine gute Entscheidung ist oder nicht ist da erstmal irrelevant, aber wenn dieser Name kommt, Fridays for Future, ich mache eine Aktion zu irgendwas, was wir potenziell falsch gemacht haben, dann geht da alles schneller. Damit kann man eben auch viel mehr Druck hinter Aktionen setzen" (N1, Appendix VII, p. 103, 35–38).

This efficacy was one of his two main reasons for joining FFF. It is the only group he really heard of before becoming involved in activism and feels it is an effective tool to excerpt pressure on politics. This is only helped by the good press around the group and spokespeople like Greta Thunberg. FFF, especially in 2019, had a lot of momentum, a lot of participants and a seemingly very positive public image (Laville et al., 2019). It was often perceived as a strong and effective movement by people like N2 and N4, acting as a reason to participate. Many individuals who were involved in other groups, also participated in FFF, as it was simply the largest player. N4, though joining FFF partially due to the perceived efficacy, lost this belief as their political identity developed. They recount:

"Aber dann glaube ich, ab dem Punkt, wo ich weiter war, habe ich halt sehr in Frage gestellt, ob es reicht und ob es dann überhaupt lohnt, diese Aktion zu machen" (N4, Appendix X, p. 120, 25–27).

After deciding FFF is not going far enough, they also asked themselves if it was even worth participating. This focus on if an action is worth it does reveal the importance efficacy has for them and could similarly have for others.

Individual and participatory efficacy
The topic of individual efficacy was addressed rather narrowly, while participatory efficacy was more prominent. Individual efficacy was mainly a factor that was lacking and therefore brought individuals to cluster into groups. XR1 puts it simply:

"Ja, also allein (...) Dinge machen bringt halt nichts. Man braucht immer andere, die da auch mitziehen und die gleichen Werte teilen. Und ja, mir war einfach klar allein werde ich hier nichts reißen" (Appendix IV, pp. 89–90, 42–1).

She realized that she cannot start actions by herself, but rather needs a group that works together and shares her values. Participatory efficacy in reference to XR was mentioned once, by XR3:

"Ja, ich denke also bei Extinction Rebellion, das ist ja jetzt in Stuttgart auch keine riesige Gruppe, also machen da einzelne Leute schon viel aus. Jetzt speziell dort" (Appendix VI, p. 101, 7–8).

He feels that his participation is significant, as the local XR group is relatively small. While this did not seem like a major influence in his decision to join, it could have had a positive impact. Within the FFF participants, N1 and N3 explained their engagement with the group and how their participation felt efficacious. N3 specifically talked about being able to get involved in the group quickly and take over responsibilities, which she enjoyed. And N1, in reference to his involvement in FFF notes:

"Deshalb, ich merke auf jeden Fall, dass ich sehr gut was verändern kann und auch schnell wichtige Sachen lernen kann" (Appendix VII, p. 105, 4–6).

Realizing that one's participation directly impacts the group and leads to positive changes could, as such, be a strong motivator to stay with the group. Taking on responsibilities may play into this as well.

The last consideration for participatory efficacy concerns N2, who ventured down the political path. For him, participatory efficacy seemed to be the most significant reason to make this decision. Many opportunities are present for him in this political realm, as he lists:

> "Kann mich […] auch thematisch wieder einbringen und habe zum Beispiel die Möglichkeit, auf einem Parteitag für den Kohleausstieg 2030 zu stimmen, habe die Möglichkeit als Grüne Jugend, also direkt als Landesvorstand, Thematiken in dem Wahlprogramm reinzuschreiben, in ein Sondierungspapier reinzuschreiben und diese Thematiken am Ende auch aus dem Koalitionsvertrag herauszulesen, die ich von der Straße aus in dem direkten Sinne aus einer, ich sage jetzt mal, Personengruppe von vielleicht 20 Leuten heraus nicht hätte" (N2, Appendix VIII, p. 110, 18–24).

He compares the efficacy of him personally in these two situations—the direct electoral political involvement and the actions on the street. He feels both are valuable, but he as an individual can enact more by participating in party politics. Recounting his decision:

> "Und deswegen habe ich den Mehrwert für mich in dem parteipolitischen Weg und auch in der Arbeit in der Partei und in einem Parlament, wo ich ja jetzt auch arbeite, entsprechend als einen größeren Mehrwert gesehen als meine Stimme auf der Straße. Das ist vielleicht der Grund, warum ich mich dann vom Aktivismus, nicht abgewendet habe, aber einfach die Priorisierung anders gesetzt habe und somit im aktivistischen Sinne nicht mehr sehr aktiv bin" (N2, Appendix VIII, p. 109, 36–41).

His decision was largely determined by the participatory efficacy he felt was stronger in the political realm. This is arguably the biggest finding for this code.

5.8   Emotions

Injustice and moral conviction

The perception of injustice or moral conviction moving people to act is both an emotion and an appraisal of an issue. It is a topic that was addressed by all people interviewed in various ways. While it is much more of an emotional moving point for some, it can also be seen as a more 'logical' moral stance. XR1 clearly states:

> "Also das Gefühl, dass mich am meisten umgetrieben hat war eben die unfassbare Ungerechtigkeit, die quasi aus meiner Sicht in diesem Kollaps oder der Zerstörung unserer Biosphäre, ja, gipfelt. Das hat sich nicht wirklich geändert" (Appendix IV, p. 92, 4–6).

This was precisely the emotion that motivated their actions and allowed them to persevere. The perceived injustice was clearly a strong influence. XR2 emphasizes their moral standpoint:

> "Ja, wahrscheinlich schon ein bisschen dieses ,ich sollte Verantwortung übernehmen' oder mehr Verantwortung übernehmen oder mich auch dafür einsetzen, für die Werte, die ich vielleicht habe" (Appendix V, p. 97, 16–18).

They feel the moral conviction to stand up for their values. This indicates a past appraisal of the environmental issues and how they are created, which go against those values. Thinking back on her emotions, N3 recounts:

> "Also es war glaube ich auf jeden Fall die Frust. Also kein ,Oh mein Gott, hey, mega-cool, Aktivismus. Mach ich jetzt mal', sondern eher die Frust ,Hey, gerade läuft so viel schief und irgendwie muss was geändert werden'" (Appendix IX, p. 116, 21–24).

She saw things going wrong in the world and assessed these as issues that needed solving. XR3 more indirectly articulates:

> "Und wenn man sich halt damit beschäftigt, dann sieht man, was alles passieren müsste, also das macht es dann immer schlimmer. Und dann dachte ich, ich probiere es zumindest" (Appendix VI, p. 99, 35–37).

He is also evidently appraising environmental changes as issues and perceives them as a worthy cause to try to fight for. This same environmental appraisal can be seen in N2's narration:

> "Also für mich ist einfach politisch die größten Krisen unserer Zeit ist die Klimakatastrophe und die Biodiversitätskrise. Das hat mich auch in gewisser Weise politisiert, also ich bin über diese Themen überhaupt politisch aktiv geworden" (Appendix VIII, p. 108, 38–40).

This appraisal and the emotions that came with it were a politicizing factor in his life. They were strong motivators of action. He specifies:

> "Und da ist natürlich ein systemisches Ungleichgewicht, dass ich mir wegwünsche und entsprechend eigentlich systemisch auch was anders wünsche. Im jetzigen System, um auf dieses Ungleichgewicht aufmerksam zu machen, um auf diese Krisen aufmerksam zu machen, sehe ich, dass man in gewisser Weise den Weg des Systems teilweise mitgehen muss" (N2, Appendix VIII, p. 110, 9–13).

The implied perceived injustices are attributed to systemic issues that he wants to ultimately change. N4 does not voice this explicitly and rather talks about their rare emotional reactions to climate events, remembering only two or three events that moved them. This perception of injustice and moral conviction seemed to be more of an underlying factor for them and not a main motivation to join activism. N1 focuses explicitly on the moral justification, succinctly stating:

> "Das ist einfach das objektiv Richtige und das mache ich" (Appendix VII, p. 106, 15).

He relays that he does not feel very emotionally involved in the issues and sees it more as a logical decision based on what is morally right.

Group emotions
Group interaction can both help alleviate emotions and can shape shared emotions. While emotions like powerlessness and grief may lead to action, XR1 asserts these can change when individuals find a group and people that feel the same, as they can support each other and feel validated. XR places a focus on this shared emotional support. She explains:

> "Und ja, durch die Art, wie XR Aktivismus machen, selber auch, also das schon immer ein großes Thema gewesen wie wir mit diesen Gefühlen umgehen, und wie wir das eher in Aktionismus wandeln anstatt das es uns nur lähmt" (XR1, Appendix IV, p. 92, 9–11).

As emotions related to climate issues can be quite impactful, they work together to process them and turn them into action rather than inaction. This focus of the group can be seen in the resilience meeting mentioned before by XR2, where exactly this processing and strengthening each other was the focus. They perceived this meeting as very positive, and it likely supported their choice to keep participating in the group. N3 remembers a situation where FFF members shared their feelings and formed opinions together. She retells:

> "Tatsächlich haben wir, glaube ich, nach dem Global Strike ganz kurz über Ende Gelände, Letzte Generation und XR gesprochen. Und da waren ja auch die Fragen ‚Hey, warum machen wir da nicht mit? Warum machen wir nicht lieber auch vielleicht mal gemeinsam was?' Und da war auch einerseits schon ein bisschen diese Angst. ‚Na ja, irgendwie müssen wir uns schon davon ein bisschen abstrahieren. Also von wegen, die sind ein bisschen radikaler, bisschen progressiver und alles.' Und da war eine gewisse Angst da, aber keine Abneigung, hatte ich so das Gefühl" (N3, Appendix IX, p. 115, 30–36).

This could indicate the group's hesitance being a factor that could hold individuals back from joining in more radical actions. Lastly, N4 recounts being moved with group actions. They say:

> "Ja. Also auf jeden Fall so eine Demo oder eine Aktion, wo die Stimmung richtig cool ist. Wenn es richtig euphorisch ist oder auf der anderen Seite richtig wütend. Das reißt einen schon mit, klar" (N4, Appendix X, p. 121, 41–43).

These shared emotions not only motivate future actions but could also strengthen the connection to the group and the people in it.

Anger and frustration

These two negative emotions seemed to strongly motivate individuals' decision to join activist groups and can be a shared emotion, as during demonstrations. N3 names frustration as the main motivating emotion to become active in activism. She recounts being told to just accept her circumstances and ultimately reached a breaking point, where she wanted to try to change those circumstances. She narrates:

> "Also, ich bin jetzt 21 Jahre alt. Ich kam vor circa zehn Jahren ja erst nach Deutschland. Ich war elf, zwölf Jahre alt circa und da habe ich schon, also natürlich nicht in breiten Massen oder so, aber bisschen auch mit Diskriminierung und Rassismus zu kämpfen gehabt. Ähm, und immer wenn ich irgendwie mit meinen Eltern drüber gesprochen habe oder so, kam da so: ‚Ja, das ist halt jetzt so. Da kannst du nichts ändern. Das ist halt jetzt so. Augen zu und durch'. Dann später, mit der Zeit, wenn man irgendwie auf Instagram oder so, oder vielleicht mal sogar die Zeitung oder sonst was gelesen hat und Sachen mitbekommen hat und mit Menschen darüber sprechen wollte, kam da auch mal so eins ‚Ja, aber das ist halt die Politik und so und da kannst du nichts dran ändern. Ist halt so. Augen zu und durch'. Und das war dann schon so ein Ding. Also Frust, auf jeden Fall. Dieses okay, irgendwie sagen die mir da, dass ich nichts ändern kann, aber ich möchte unbedingt was ändern" (N3, Appendix IX, p. 116, 8–18).

Her built-up frustration motivated her to act and become politicized. A similar frustration about the lack of action is named by XR3:

> "Ich denke der Antrieb ist so, dass man sich halt denkt, gerade wenn man sich viel mit dem Thema beschäftigt und sich dann auch anschaut was passiert und halt eben nicht passiert, dass das einen dann ziemlich frustriert auch, dass eben nichts passiert. Und wenn man sich halt damit beschäftigt, dann sieht man, was alles passieren müsste, also

das macht es dann immer schlimmer. Und dann dachte ich, ich probiere es zumindest"
(Appendix VI, p. 99, 32–37).

A frustration about nothing happening can develop, intensify, and thus motivate
action. He continues:

> "Und wenn man jetzt bei Extinction Rebellion so, ist bisschen mehr das Gefühl da,
> von ich mach was dafür. Eben dieses Versuchen. Und was dann schon auch diese Frus-
> tration ein Stück weit nimmt, weil man eben nicht mehr ganz so untätig dasitzt und
> ja" (XR3, Appendix VI, p. 101, 21–24).

The frustration that motivated action can be alleviated through the feeling of
doing something, of agency, and of understanding in the group. Anger is one of
the few emotions mentioned by N1, stating:

> "Ja, über Inkompetenz von Politik rege ich mich immer auf, das haben wir vorher
> schon gemacht, aber das ist auch mein großes Ding. Und ansonsten gehe ich da relativ
> nüchtern dran" (Appendix VII, p. 106, 25–27).

Even though he says that he does not feel emotionally invested, this anger could
still be a factor that may play into his activism. He agrees that involvement
in activism helps lighten some of the frustration. Anger is also one of the few
emotions mentioned by N4 in combination with group emotions, though they feel
that it might not be as extreme as for others. They comment:

> "Wenn es richtig euphorisches oder auf der anderen Seite richtig wütend. Das reißt
> einen schon mit, klar. Aber ich kann mir vorstellen, dass es bei mir schon weniger ist
> als bei anderen. Aber ich weiß es nicht" (N4, Appendix X, p. 121, 42–44).

Guilt
While XR2 and N1 mentioned some feelings of guilt that might have motivated their actions, this was not a prominent topic found in the interviews. XR2
comments:

> "So ein bisschen Schuldgefühle oder ein schlechtes Gewissen kann ich mir schon
> vorstellen" (Appendix V, p. 99, 15–16)

This comment was in response to a directed question, which indicates its limited
importance.

Agency and powerlessness

A sense of being active and able to enact change is a feeling that individuals can be in search of when joining activist groups. As seen in previous sections, this sense of agency is effective in alleviating negative emotions like anger and frustration. XR1 explains:

> "Es spielen ja auch andere Gefühle eine Rolle, wie, sagen wir Trauer und Ohnmächtigkeit, weil man denkt sich nichts ändern zu können. Und ja, das ändert sich auf jeden Fall, wenn man dann eine Gruppe von Menschen hat, die das ähnlich sieht" (Appendix IV, p. 92, 6–9).

Powerlessness and grief were emotions she mentioned originating due to feeling like one cannot change anything. Through identification with a group that may share those feelings and engage in actions, they are eased. XR1 recounts that this hope for agency through a group allowed her to persevere through a slow start of the group. This sense of agency takes precedence over efficacy beliefs. As XR3 emphasized:

> "Und wenn man jetzt bei Extinction Rebellion so, ist bisschen mehr das Gefühl da, von ich mach was dafür. Eben dieses Versuchen" (Appendix VI, p. 101, 21–22).

Though he is not sure of the group's efficacy, he highly values the action of doing something, of trying.

This sense of agency can also be seen more generally to motivate activism. N2 was motivated by possible agency through FFF as well, considering:

> "Das heißt, ich kann mich zwar persönlich einschränken, aber damit wir wirklich global was ändern, muss das ganze systemisch passieren, das heißt, das muss politisch passieren, es muss gesellschaftlich passieren und dafür ist es relevant, dass man sich gesellschaftlich auch einbringt, aus meiner Sicht" (Appendix VIII, p. 109, 2–5).

He necessitates the agency through group action in society to enact change.

Hopelessness and grief

Other than XR1's mention in the previous code, grief and hopelessness were otherwise mentioned as emotions experienced by people around the interviewees and not by themselves. N4 reflecting:

> "Vor allem, weil ich da schon öfters mit Leuten hatte, die halt meinten, dass für sie so Weltschmerz und so Hoffnungslosigkeit so ein ganz, ganz großes Thema ist und dass

es sie total fertig macht mit Klimakrise und anderen Krisen in der Welt" (Appendix X, p. 121, 25–27).

A similar sentiment is shared by XR2, who mentions the grief people feel in response to the issues they see in the world. Both N4 and XR2 distance themselves from these emotions, but through their experience with other people still bring to light these topics.

5.9   Self-Identity

Politicized identity
In a way, a politicized identity is a prerequisite for involvement in activism, as activist groups are inherently political. The interviews revealed how some individuals were politicized through the climate crisis. N2's story starts off in school, he retells:

> "Also für mich ist einfach politisch die größten Krisen unserer Zeit ist die Klimakatastrophe und die Biodiversitätskrise. Das hat mich auch in gewisser Weise politisiert, also ich bin über diese Themen überhaupt politisch aktiv geworden. So ein bisschen gestartet durch schulische Bildung zu Ernährung, das heißt, mich hat es dann über den vegetarischen, veganen Zweig so ein bisschen die Thematik Klimawandel überhaupt erst erreicht" (Appendix VIII, p. 108, 38–42).

His politicized identity was formed through these issues, which he was made aware of initially through school education on nutrition. N1 was only confronted with climate topics later. However, after informing himself further on the issues, his previous apoliticality quickly turned into a politicized identity, as he decided to act. N3 follows a similar initial politicization as N2 in school, though her school was more actively engaged in social action. This social engagement combined with her urge to enact change, allowed her to find her agency in political action. N4 recounts how they politically developed only through their engagement in FFF, moving on to more radical movements from there. Finally, XR2 describes their politicization through a realization that they wanted to stand up for what they believe in. They describe:

> "Ja, ich glaube bisher war es schon öfters so, dass ich quasi erwartet habe, dass irgendein Angebot da ist und ich das dann annehmen kann. Und das war schon irgendwann so ein aha-Moment glaube ich, dass ich gemerkt habe, ich kann auch selber was

dafür unternehmen. Zum Beispiel auf Menschen zugehen oder zum Beispiel eben irgendeine Form von Aktivismus machen" (XR2, Appendix V, p. 97, 20–24).

This development allowed them to become more politically engaged.

Support of self-image

Participation in activism can be a way of confirming one's self-image. N3, for example, started off becoming socially engaged in school and has since kept participating in various groups. This indicates a certain commitment to oneself in that social engagement. One of the most named reasons for joining XR in the open meeting was also named to be previous engagement in activism. This signifies a commitment to supporting an activist image of oneself. N4, on the other hand, saw themselves as an *'Ökokind'*, a child that grew up in an environment that is already sustainably conscious and might have already engaged in topics like the climate crisis. They phrase it as:

> "Aber sonst war das eher so ein Ding von, ich war schon immer so ein Ökokind und irgendwie ich war der komische politische Mensch in der Schule. Und dann war das nur so der logische nächste Schritt. Dann irgendwie freitags auf die Straße zu gehen, weil ja, was will ich sonst machen?" (N4, Appendix X, p. 121, 32–35).

It seemed logical to them to start demonstrating with FFF, as it fit with their image of themselves.

Identity development

The development of one's own identity was named as a prominent reason by XR2. They state:

> "Also, ich glaube ein wichtiger Grund war für mich, dass ich mich selber verändern und weiterentwickeln wollte" (XR2, Appendix V, p. 94, 26–27).

While this was not a widely mentioned topic, it was an important reason for them. N3's feeling of powerlessness that developed into a decision where she decided to try to change things can also fall under this code, as she took action, stepped out of her comfort zone, and developed significantly in her identity.

5.10   Additional Codes

Responsibility
This injunctive code played various roles in the decision-making processes of the individuals, though mostly as an interacting factor rather than a direct influence. XR2 considers:

> "Ja, ein Punkt ist noch so ein Punkt Gruppenstruktur und Verantwortung übernehmen. Also es gibt manche Gruppen, die sind halt klar hierarchisch organisiert [...]. Was den Vorteil hat, dass klar ist es gibt ein Programm und man kann einfach dazukommen und mitmachen und bei XR ist es dann manchmal ein bisschen anstrengender, weil man eben selber sich auch beteiligen muss, manchmal, weil sonst eben manche Sachen nicht übernommen werden oder weil da halt alle ein bisschen in der Verantwortung sind. Was aber wenn man sich da so ein bisschen reingefunden hat auch schön ist, zum einen eben zu lernen selber Verantwortung zu übernehmen und zum anderen bedeutet das dann auch sich verantwortlich zu fühlen und es ist dann halt mehr so ein Commitment, das heißt es ist auch dann schwieriger ein Stück weit sich wieder rauszuziehen" (Appendix V, p. 95, 5–15).

They explain that XR can create different experiences with responsibility considering their non-hierarchical structure. This means that the individuals involved all carry some of the responsibilities. They evaluate this structure to be more demanding but also more rewarding. While participants are compelled to commit more, they also have more of a say and can learn how to handle the responsibility. It can, however, be more difficult to pull back from the group when taking on this responsibility. While many said they liked being able to take on responsibility in FFF as well, this was also a limiting factor in engaging with other groups like Extinction Rebellion. N3 mentioned how the responsibilities she was able to take on quickly allowed her to feel more like part of the group, but she also says:

> "Wie gesagt, also ich würde schon sagen, was mich einerseits davon abhält, bei XR mitzumachen, ist halt, dass ich derzeit ziemlich wenig Kapazitäten habe, dadurch, dass ich mich in anderen Bereichen viel mehr engagiere" (Appendix IX, p. 114, 19–22).

Her engagement and duties in other groups hold her back from taking the time to participate in XR. This reason was named by others too. N1 explains:

> "Dann kommen eben die ganzen außenstehenden Gruppen mit Letzte Generation oder Animal Rebellion, Peta, (unv.), sowas was es eben alles gibt, und ich glaube das Problem ist vor allem, wenn man erst so spät von den Gruppen erfährt, dann hat man ja

irgendwie schon seine Gruppe, wo man was macht, und steckt da vielleicht auch seine Zeit rein und kommt dann eher selten dazu bei was anderes noch was zu machen" (Appendix VII, p. 104, 5–9).

The responsibilities he was happy to take on in FFF also hold him back from being able to spend time in other groups. N4 similarly mentions:

"Ganz abgesehen davon, dass ich einfach schon anderweitig organisiert war. Also, dass ich einfach schon bei FFF war, dass ich irgendwie in Stuttgart andere Sachen gemacht habe" (Appendix X, p. 119, 9–11).

This being only a small part of the reason they did not join XR, the bad image they had of the group being more significant. Responsibility could interact with group identification and participatory efficacy factors. Group identification can be higher when more engaged with the group and participatory efficacy can similarly be increased.

Image

The way a group is perceived by an individual, their social circle, or the public proved to be a noteworthy topic. Generally, a perceived negative public image did not speak against joining XR, unless that image pertained specifically to norms and values. Positive and negative images from closer social circles, however, strongly influenced opinions and decisions to participate. The lack of an image of XR also negatively influenced participation. N4 who remembers only finding out about XR through friends and fellow activists is likely to have gained a negative image through those circles. They explain:

"Eigentlich mit so dem Image, das XR hat bei Leuten auch in der Klimabewegung, die XR nicht so geil finden. Es auch noch mehr so dieses, dass es irgendwie auch viel so Antisemitismusvorwürfe gibt gegen einzelne Personen oder viele Leute aus der Bewegung oft so ein unreflektierter Umgang mit irgendwie Rassismus und solchen Sachen sind halt oft" (N4, Appendix X, p. 122, 5–9).

And:

"Ja. Also, in so den Kreisen mit den engsten Genoss*innen, mit denen ich darüber geredet habe oder wo das Thema war, war es schon eher so, dieses sich über XR lustig machen und sagen, das sind ja die, weiß nicht, die Esofritzen aus der Klimabewegung so bisschen" (N4, Appendix X, p. 121, 16–19).

Not only does this negative image go against N4's norms and morals, as it is specifically related to sensitive topics like racism, but it is also shared with close friends and activist acquaintances. This allows it to be reproduced and strengthened. And an activist group that is made fun of amongst friends is unlikely to be one they would then join. Less closely produced perceptions, however, seem to play less of a role. N2 makes clear:

> "Aber die Wahrnehmung bei Extinction Rebellion hindert mich nicht daran, oder die von manchen politischen Gruppen vor allem kommt, hindert mich nicht daran, da mich zu beteiligen" (Appendix VIII, p. 111, 13–15).

The image that the public or political groups might have of XR is not what is stopping him from participating in XR. N3, though mainly off-put by a lack of an image of XR, supports N2's standpoint:

> "Was ich schade finde, ist halt, dass vor allem in den deutschen Medien XR ins Schlechte gezogen wird. [...] Aber das sind auf gar keinen Fall Gründe, die mich davon abhalten würden, da mitzumachen" (N3, Appendix IX, p. 114, 14–19).

A bad image from the public is not a factor that would stop her from joining the group. She also feels like there is more of a lack of an image of the group which ties back to her perceived lack of efficacy. N1, though friends with XR participants, likewise feels more of a lack of an image, while XR2 received a very positive portrayal of the group from connections to individuals in the group. N1 also claims to have never heard a word against XR from people in FFF, which is contradictory to statements made by N4. This could indicate differing opinions or changes in perception over time, as the two could have been active during different times. While XR3 did feel there was a negative image of XR in the media, this did not stop him from joining the group.

5.11   Summary of the Results

The present research has shown various patterns in the decision-making process of participating in XR. Factors include but are not limited to social identity, norms and morals, efficacy, and emotions, as seen in the literature. The results can be consolidated into the following initial observations:

- Voiced on several occasions, individuals value a group setting that makes them feel it is 'their group'. Group identification plays a role in choosing and sticking with an action group.
- A lot of people join groups, where they already know people and have a connection to them, though knowing people in a group is not a standalone predictor of participation.
- Strong social connections are formed within action groups, which can be a feature an individual seeks out or it can be a reason to remain with a group.
- Outside influences played less of a role or a less visible role in deciding to join activist groups. Parents with more critical views might dampen the level of action an individual ends up engaging in.
- Group dynamics, which could influence group identification, are a factor that could sway the decision to join a specific activist group.
- Moral beliefs and values generally can influence an individual's impression of an activist group.
- XR's actions, like their civil disobedience, are generally seen by activists in a positive light, sometimes motivating participation, especially due to the flexibility of which actions an individual wants to join.
- The perception of the descriptive and injunctive norms of an activist group and the agreement thereof can be a determining factor in joining an activist group.
- While collective efficacy did not play a determining role in participation for XR members, it seemed to play a significant role for those involved in FFF.
- Participative efficacy more so than individual efficacy can play a role in the decision to join an activist group.
- Recognizing the climate crisis as a perceived injustice and finding fighting against it necessary and 'the right thing to do' is an almost universal prerequisite for activist involvement.
- Shared emotions can motivate joining a group, as well as originate from acting or processing together.
- Anger, frustration, guilt, hopelessness, powerlessness, and grief are all emotions that can contribute to a motivation to seek out groups that engage in activism, which can help alleviate those emotions. They were, however, seldomly a significant talking point in what determined individuals' decisions.
- Engaging in activism can give people a sense of agency, which they might lack otherwise.

- Engagement in activism can be a way to develop one's identity or a way to reinforce one's self-image. A past in activism was named by several people as a determining factor in their decision to come to XR.
- Being able to take on responsibility in an activist group can be a positive factor in joining and can also prevent activists from joining other groups.
- XR's image or perceived image can be a determining factor in the decision of individuals to participate.

Discussion

6

The results present a wide variety of findings, which further support the relevance of additional research in the environmental activist realm. The following chapter aims to bring these extensive results into the context of the theoretical background, specifically discussing patterns, model fits, and the significance of social identity. During this discussion, it is pertinent to keep in mind the nature of the results, as they are case-specific, defined by the local Stuttgart XR group, and solely activists' or past activists' perspectives. The research questions hereby remain at the forefront. To reiterate, these are:

- What patterns can be identified in the decision-making process of people joining the environmental activist group Extinction Rebellion in Stuttgart, Germany?
- To what extent are the existent models SIMCA, EMSICA, and SIMPEA applicable in this environmental activist realm?
- How does the significance of social identity manifest itself in the decision-making process?

In an attempt to make the existent patterns more apparent, the chapter will start by considering visualizations of the individual's decision influences. Furthermore, these patterns will be discussed as a whole. In order to tie this back to the theoretical background, the fit of the SIMCA, EMSICA, and SIMPEA models will

Supplementary Information The online version contains supplementary material available at https://doi.org/10.1007/978-3-658-44047-3_6.

Y. Plate, *Social Identity Motivators in Environmental Collective Action*,
BestMasters, https://doi.org/10.1007/978-3-658-44047-3_6

be evaluated in combination with further theoretical considerations. The discussion will be closed off with considerations of the research's limitations and future research opportunities that present themselves.

6.1   Visualizations of Individuals' Decisions

The following visualizations in Figures 6.1–6.7 provide a simplified overview of the various influences on the individuals' decisions. Social identity is posed centrally, as it allows individuals to see themselves as a part of a group. Participation in an activist group like FFF and XR presupposes collective action. The groups' main purpose is collective action and the joining of one of those groups implies the involvement in this action. This makes social identity and collective action almost synonymous. While they are two separate entities, in this context they are inherently intertwined. Looking outside of the activism group context, individuals who are not for example active members of FFF, but still participate in their demonstrations would present involvement in collective action without a strong social identity with the FFF group. This is less likely to occur for XR, as the more radical actions require some training and preparation beforehand. Social identity without collective action—in this context—is rather unlikely, as both XR and FFF are politicized groups. The identity content—the content of the XR or FFF identity—is action focused and politicized. Joining this group would suggest an intention of participating in collective action, which was the case for all interviewees. This strength of the identity content supports the value it was given in the extended SIMCA (van Zomeren et al., 2018).

The social identities in the visualizations (Fig. 6.1 & 6.3–6.7) are all politicized, as these are identifications with the two activist groups. One slight exception should be mentioned, as N2 (Fig. 6.2) chose the electoral political path (no social identity, but still politicized) and therefore no longer engages in collective action. Nevertheless, the social identity was still included, as there were aspects that in the past motivated him to join activism groups. For the rest of the interviewees, as the activist groups are politically charged and clearly stand and act against what they perceive as unjust—the climate crisis and its handling—joining these also necessitates the agreement with that perceived injustice[1] and a desire to act against that perceived injustice. As such a necessity, perceived injustice is present in all visualizations (Fig. 6.1–6.7), as a motivator for

---

[1] The perceived injustice term here acts as an umbrella term, also pertaining to a perceived environmental threat or a feeling of moral conviction.

collective action and identifying with a social identity. For the interviewees, it seemed to both directly motivate action and to motivate joining a specific group through which they could act. As this group is only sought out after identifying a perceived injustice, the connection between injustice and social identity follows EMSICA's route. As can be seen in the visualizations, perceived injustice motivated social identity and collective action. This does not rule out that this perception was strengthened, supported, or encouraged through social identification with the groups. Generally, the factors can be seen to influence social identity, which seemed to be the main relation and was the focus of the research. Nevertheless, social identity can still influence factors, such as emotions, or could be a platform for new social connections. While most factors seen in the visualizations influence social identity, social identity can also reflect back onto some factors. For XR1 (Fig. 6.5), for example, the relationship between emotions and social identity goes both ways.

The visualizations make evident the varying importance that factors play for different individuals. While participatory efficacy seems almost solely responsible for N2's decision (Fig. 6.2), activists involved in FFF (N1, N3, N4) have several factors that spoke in favor of FFF rather than XR (Fig. 6.1, 6.3, 6.4; marked with FFF on the arrow). XR3 (Fig. 6.7), who said image was a point speaking against joining XR, did not view this as relevant enough to determine his decision. Such influences speaking against joining XR are marked with an X on the arrow. Social connection, while not ensuring a decision towards a group, could be determining especially through a perceived image of the group. Social connections to both groups were present for N1 (Fig. 6.1) and N2 (Fig. 6.2), who both put more weight on efficacy for their decisions, thus joining FFF. While N4 (Fig. 6.4) had social connections to XR, other social connections conveyed a negative image, which strongly influenced N4 against joining XR. On the contrary, the positive image conveyed by XR2's social connections strongly influenced them into joining XR. A factor that was a topic for everyone was norms and morals, though less directly for N1. A dotted line for that influence (Fig. 6.1) communicates a slight or implied influence. The visualizations will not include all influences individuals had in their decisions, as there are likely ones that were not mentioned explicitly. Significant ones which did not play a role are included underneath the individual models for easier comparison. Overall, the strength of the influences on the decision to join XR vary greatly between individuals. Nevertheless, initial patterns concerning norms and morals, perceived injustices, social connections, and efficacy perceptions can be identified. This will be further elaborated on in the following section (Fig. 6.6).

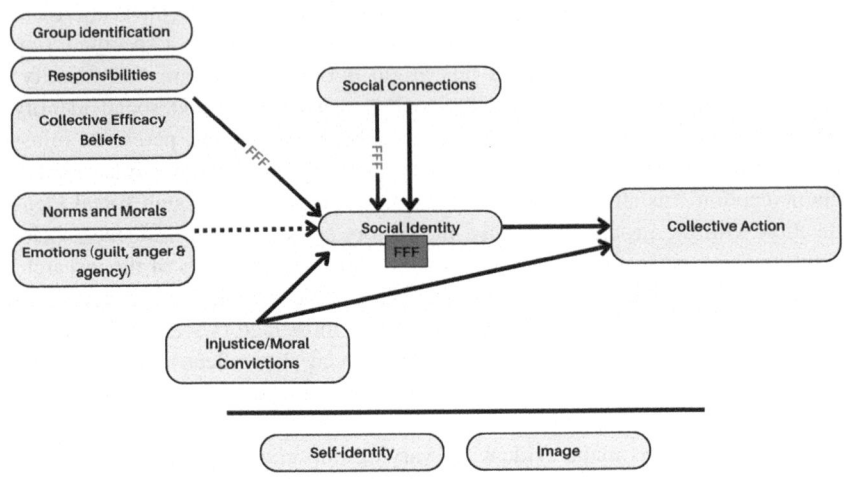

Figure 6.1 N1 Visualization

6.2   Patterns and Reflections

The deductive and inductive analysis of the interview transcripts offered both support of aspects of existent models and factors and provided contradictory views, further extension possibilities, and enrichment of the theory by investigating the existent models in a new realm. The various factors and their weights for the interviewees provided a plethora of results that require further discussion to contextualize them in the research and theory. This will follow the categories, as used in the results.

As predicted, social identity holds its place at the center of the factors influencing the decision to join a collective action group. Social identity, as individuals' connection to the social world, proved to be extremely relevant. Its relationality can be seen through social connections and shared norms and its shared and meaningful nature is apparent through the shared experiences and their strengthening effect on the identity. This strong core holds its place in the center of collective action motivators. Social identity as a factor proved to be more multifaceted than expected based on the existing literature. Especially the effect of inter and intra-group social connections. The social identity that individuals adhered to is formed through several factors. Group identification, as a prominent theoretical aspect of social identity, remained relevant, though often more subtly

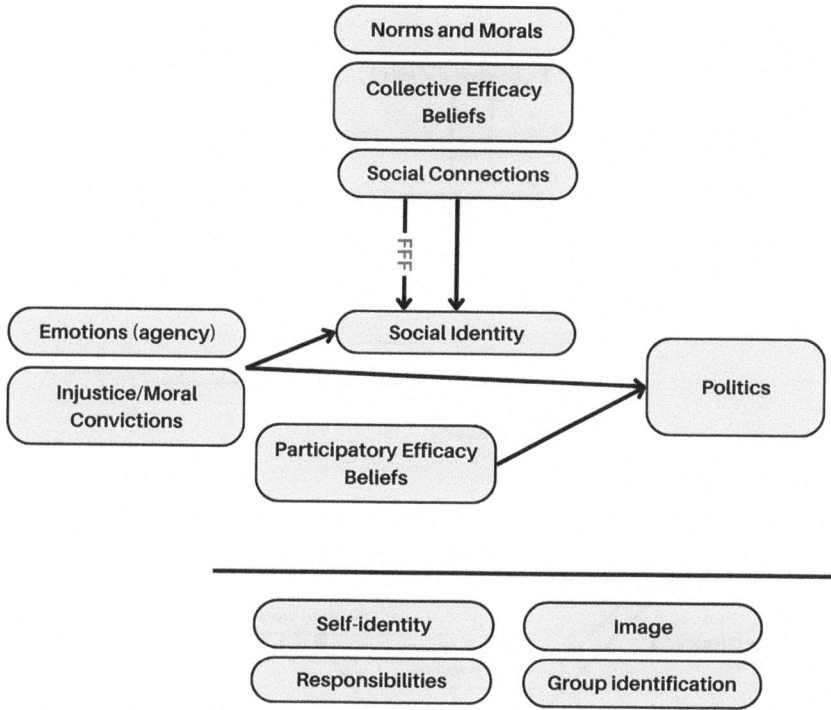

Figure 6.2 N2 Visualization

or in the form of a feeling. Rather than a strong identification with a group, individuals more often simply stated their involvement and their role(s) within a group. This general identification with a group was aided by agreement with the group's characteristics, goals, norms, and actions. Agreement with a group's norms and morals, especially for XR members, was a prominent way that people related to their group. This follows the relationality of social identity (SIA; Reicher et al., 2010), where individuals define themselves in terms of their similarities and differences in relation to others. Further, it falls in line with normative fit categorization (as suggested by van Zomeren et al., 2018), where identification follows normative alignment. The agreement with the group's norms and morals is a significant factor to find identification with. Relationality further connects to social connections. Finding similarities with people within a group in comparison

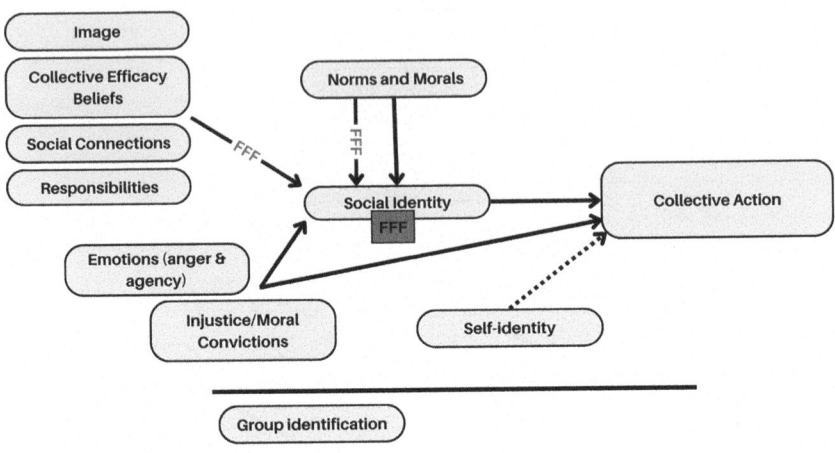

Figure 6.3 N3 Visualization

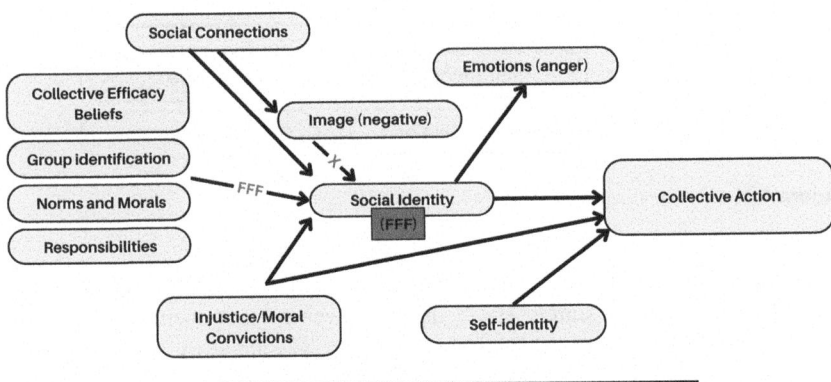

Figure 6.4 N4 Visualization

to differences with people outside of a group gives a strong argument for identification with that group. This could occur once already a part of a group or it could be a factor that convinces individuals to join a group. If they have social connections that they relate to, who are already a part of the group, individuals are more

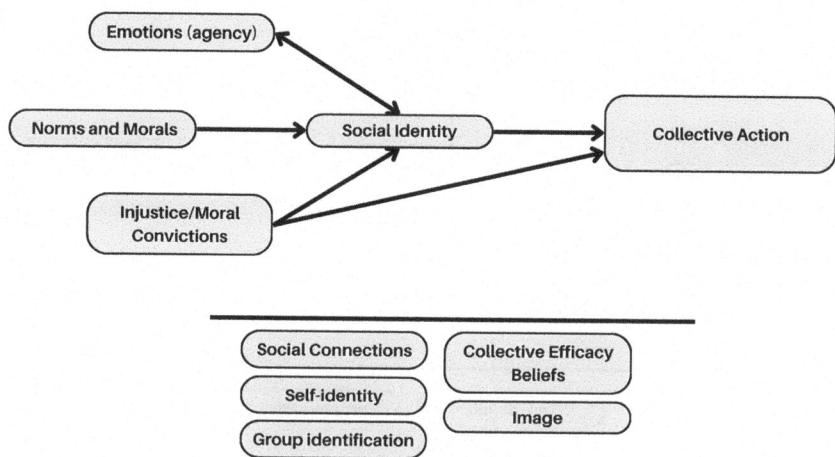

Figure 6.5 XR1 Visualization

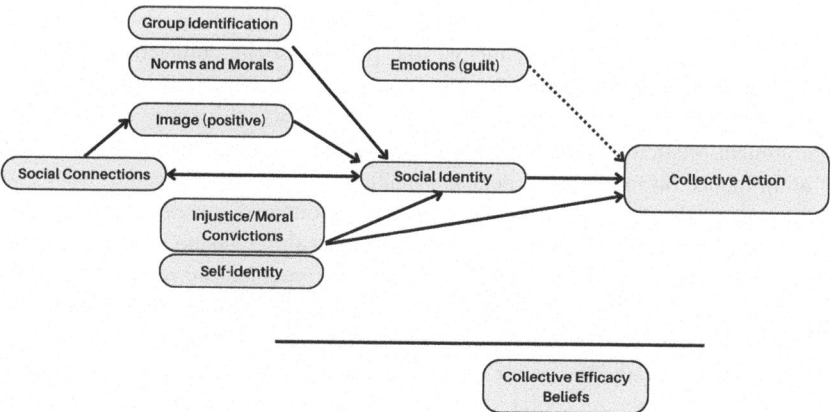

Figure 6.6 XR2 Visualization

likely to consider joining the group. This pattern could be seen on several occasions throughout the research. Social connection, through its social nature, very clearly also plays into social identity. It can be an influence to bring people into

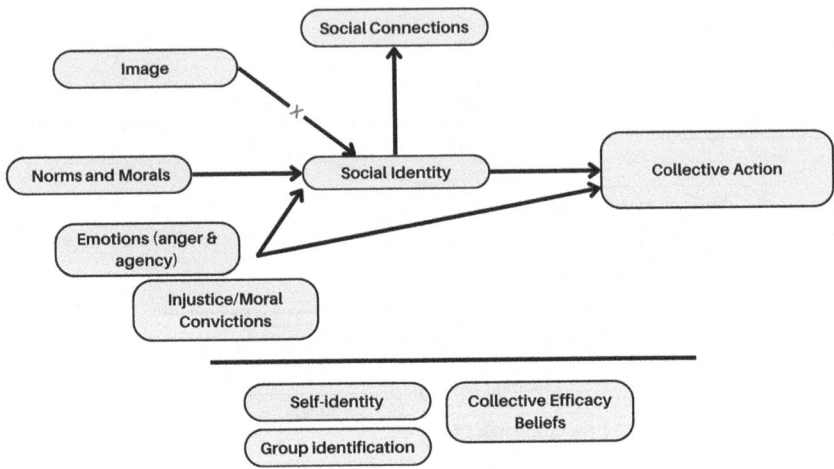

Figure 6.7 XR3 Visualization

the group and it can strengthen social identity within the group. Furthermore,—
though this is a less significant factor—the possible social connections within a
group could be a desirable point, which could positively influence an individual
to join. Overall, social connections and their role in social identity proved to be
a prominent predictor. Outside social influences on the other hand had much less
of an influence on individuals' decision-making process than expected. Rees and
Bamberg's (2014) findings on perceived participation norms, seemed to only have
limited application in the environmental activism realm. Outside influences could
have slightly reduced the radicality of the groups joined but did not have a strong
influence overall. This does not, however, exclude possible influences on individuals who are not involved in activism at all. Here this factor may play a more
significant role. Lastly, the factor of group dynamics and how they are perceived
can have an influence on social identity. This newly added factor was mentioned
by several individuals and seemed to play a role in how individuals saw the
groups. It could be a factor in identification with a group and perceived norms. If
a group's dynamic speaks to an individual and seems to align with their norms,
they are more likely to feel comfortable joining that group. Seeing issues with a
group's dynamic could point to disagreement with certain morals that pertain to
things like power dynamics and group structures. This could discourage further
involvement. Group dynamics is a factor that is susceptible to perceptions and

brings a new dimension into what can affect social identification. Social identity with its various facets remains a vital factor to consider when it comes to joining environmental collective action groups.

The second extremely significant factor in this investigation was norms and morals. Both can play a determining role and their relevance in previous models and research is confirmed with this research. It seemed especially relevant when deciding which group to join. As one of the first things mentioned by all three XR members, norms and morals were one of the most prominent factors to influence decisions to join XR. Closely followed by group identification and injustice perceptions, which were mentioned by some XR members in their initial narration as well. The importance of norms and morals was not limited to XR members and was also discussed substantially by the other interviewees. The development of norms and morals from a social identity could be seen less in this research. This could partially be drawn back to the lack of research focus on the outcomes of group involvement. Nevertheless, the political development through group involvement, as explained by N4 could point towards such influences. N4 also brings to light the role perceptions play here, as they perceived the norms (especially injunctive) and morals very differently from XR members. These biased perceptions likely exist on both sides, aided by confirmation biases. An unexpected finding was the consensus of support for the more radical actions of XR—in the line of civil disobedience—which was expected to be a more controversial point based on public perceptions. This could be attributed to the demographic of activists in the interviewees, as they are already prone to agree with actions that aim to fight for the same cause. Nevertheless, this has indicated that XR's actions were generally supported by other activists. Norms and morals in their various iterations are vital to consider in the environmental activist realm, as they are closely connected to injustice perceptions and action intentions. The group's norms and morals must align with how the individual perceives the world and the issues that they want to solve, as well as how they want to solve them.

Efficacy proved to be a more controversial topic than anticipated, with its predictive quality seemingly far lower for XR than previous models would suggest. FFF participants valued perceived collective efficacy much higher than participants in the local XR group. Here a sense of agency seemed to play a more significant role. People involved in FFF often mentioned collective efficacy as one of their first reasons for choosing to participate in FFF over XR. A negative or lack of an image of XR was also mentioned as an initial reason, which could influence efficacy beliefs. A lack of an image of XR would not promote high collective efficacy beliefs. A possible explanation for lower efficacy beliefs is the nature of the disadvantage that is being fought. The immediate politicization of the group discussed earlier reflects what van Zomeren et al. (2008) explain happens in response to incidental disadvantage. When responding to a situational issue that is deemed unjust, the group forms around that incident to rectify it, which establishes the group as already politicized. Structural disadvantage would require a group that already exists in the social realm to become politicized to fight against a systematic issue that they need to deem unjust. Climate change and biodiversity issues manifest in incidental disasters. And while XR, as well as other environmental activist groups, were formed around the issue of the climate and biodiversity crisis they saw through incidental issues, they place a specific focus on the structural issues that ultimately caused those crises. As structural issues are much more difficult to fight and little progress has been made (in the realm of the climate crisis), this might lower collective efficacy beliefs. However, this explanation does not account for the difference in efficacy beliefs between FFF and XR. A feeling of hopelessness or powerlessness due to this structural difficulty can be turned into a feeling of agency through social identification, but it does not ensure a belief in efficacy in the case of XR. This contradicts the relation argued in the PCI model, where a sense of agency does increase efficacy beliefs (Simon & Klandermans, 2001). It can be argued that XR places more of a focus on specific structural issues than FFF, as they seem to be more set on fundamental system change. This is, however, dependent on how the groups are perceived and not a clear fact.

This structural-incidental conflict opens up a vital concern in the application of collective action models on environmental action groups. It also makes apparent the lack of consideration of such groups in the initial construction of the SIMCA, in which these different disadvantages are addressed. Though this consideration might play into the lack of perceived efficacy for XR, the strong valuation of collective efficacy in FFF members requires the consideration of additional factors to account for the difference in importance. The small size of the local XR group in comparison to the likely much larger FFF group could be another explanation for the difference in belief of efficacy. Given the size of the FFF group and their much more present and generally positive publicity, it is likely that participants will view this group as more efficacious. The difference in valuation of that group efficacy might be the discrepancy in its determining effect in the decision of which group to join. Individuals might value the identification with the group and alignment with their norms and morals higher than perceived collective efficacy. XR members explicitly said that they value their feeling of agency through activism more than perceived efficacy beliefs. This varying valuation is, thus, an important consideration. Individual and participatory efficacy was not a very prevalent or elucidating factor. Participatory efficacy proved extremely relevant for one person but was otherwise a rather insignificant factor. This can point back to the varying valuation. N2 valued participatory efficacy, which he believed was higher for him in politics rather than activism. Collective efficacy is a much more relevant factor in this research and brings about new considerations.

Injustice perceptions, moral convictions, or environmental threat perceptions were the most relevant emotional factor in motivating action and joining action groups. As a universal factor throughout the research, it could be deemed a prerequisite for joining an environmental action group. This might be said for all social identities that are involved in collective action, especially ones responding to incidental disadvantages. It is not the sole predictor of action but to get involved with a group fighting a specific issue, one has to perceive the said issue as unjust and necessary to fight for. In line with previous studies, this research found this factor to be present for all interviewed individuals in different forms. Some saw the issue as a clear injustice, others as a moral issue that needed righting, but all individuals agreed that they wanted to do something about it. Feelings of agency, which motivated and was strengthened through participation, could be connected to a feeling of efficacy, as discussed previously. In that sense, it is also a relevant emotion to consider. Guilt, anger, frustration, and other emotions which were addressed in the interviews can be especially important in climate crisis topics but did not offer clear predictive qualities from the current research. Individually, they did motivate some action, but they were never a strong reason for or against joining a collective action group. Shared emotions similarly did not provide predictive qualities but did seem an important aspect, especially for XR. They were more of an interacting factor, strengthening social connections within the group and fostering group identification. Overall, the most relevant factors pertaining to emotions remain the injustice perceptions and the sense of agency.

The newly added factors, self-identity, responsibility, and image were found to play a role in the decision-making process of individuals. The support of one's self-identity can also be supported as a factor. This supports Fielding et al.'s (2008) and Bamberg et al.'s (2015) arguments that individuals act in a way to support their self-image and are more likely to participate in collective action when they have already in the past. This previous research, supported by the present research, strengthens the argument that this should be taken into consideration in future models of environmental collective action. Self-identity is further noteworthy in its self-development, which can be connected to politicization. Politicized identities and their prerequisite role have already been discussed. Self-identity in both its self-confirmation and its politicization form can play into social identity. Responsibility as a factor can also influence group identification, as it can make individuals feel more of a part of a group. It is, however, more of an interacting factor. While it can influence people to stay with a group, most of its influence is through its interaction with group identification. Image, on the other hand, can be very prevalent and strongly influence perceptions. It can be a more vague influence through a general image an individual might have through the media, or it can be a stronger and more specific image through closer social connections. This is especially relevant for evaluating influences on decision-making. As it has become apparent that there are consistencies and discrepancies with previous models, it is now crucial to concretize these.

6.3   Model-Fit Considerations

Based on the findings from this research, it is made clear that the collective action models need adjustment and expansion in the realm of environmental activism groups. Reflecting on SIMCA, EMSICA, and SIMPEA, several aspects remain relevant, while others require adjustment to be applicable to groups like XR. Environmental activism is not fully comparable to collective action originating from previously established social groups and similarly is not fully comparable to groups forming in response to incidental disadvantage, though this comes closer. As discussed previously, environmental issues form a complex mixture of structural and incidental issues. While they can especially affect a specific group, the large-scale climate and biodiversity crisis affects humanity as a whole. The groups that form around these issues can be catalyzed through specific climate events, but act on the more all-encompassing and structural issue. This issue becomes more complicated due to the lack of a clear other to fight against. This fundamental difference in the main issue centrally sets it apart from most other social issues fought through collective action. SIMCA and EMSICA do not adequately consider these environmental collective action groups.

The social identity's position in the models follows most closely the EMSICA's causal relations. SIMCA does not consider factors that go into the identification with a social group, rather, it investigates the relation between social identity and collective action. In this model, collective action motivations mainly originate from social identity. This relation is largely irrelevant in the case of environmental action groups, as social identity presupposes an intention to act. The close connection between social identity and collective action in these groups can be reiterated. While some factors involved in collective action motivation can be strengthened through social identity, the current research indicates that for environmental issues, social identities act mainly as intermediaries. This would be more in line with EMSICA's causal relations. A social identity that is inherently connected to collective action does not leave much room for factors in between. SIMPEA places ingroup identification in the same realm of factors as collective efficacy beliefs and ingroup norms and goals. Social identity as such is not placed centrally, likely due to the inclusion of all PEB and not solely activism. This formation lacks focus. It cannot quite be applied to environmental activism groups, as many of the influences do not directly connect to social identification. Social identity groups need to remain central when looking at collective action. As mentioned by XR1: "Ja, also allein (…) Dinge machen bringt halt nichts." (XR1, ESM, Appendix IV, p. 89, 42). Doing things alone does not get you anywhere. To be able to engage in collective action, you need a group. Thus, this group needs to be viewed centrally.

Though SIMCA's initial relational structure needed to be dismissed, the idea of a twin core of (violated) moral beliefs and (politicized) identity (van Zomeren et al., 2018) should be considered. Both the politicization of the groups and the importance of norms and morals to identify with the group have been shown. As norms and morals played a significant role in individuals' identification with their groups, a close connection of the two factors can be suggested, as through normative fit categorization (based on SCT and suggested by van Zomeren et al., 2018). While EMSICA suggests the permeation of norms throughout the model by proposing a separate model, the normative alignment model, it might benefit visibility to include norms and morals, as well as their interactions more presently. In the extended SIMCA, moral conviction, which determines a cause to fight for, is included in these moral beliefs. For the investigation into environmental action groups, it will be suggested that this appraisal of the issue is kept separate. The perceived injustice, perceived environmental threat, or moral conviction seemed to play a role separately from the groups for the individuals. They motivated them to act in general, which ultimately brought them into activism and their participation in activist groups. Thus, it might be favorable to separate the issue appraisal from the norms and morals, even though they are fundamentally linked. The two factors played separate roles in this research, where issue appraisals usually motivated action more generally, while norms and morals and their alignment influenced which group an individual joined to act through. Consequently, a twin core of norms and morals and a politicized social identity can be considered. Issue appraisals, such as perceived injustice, perceived environmental threat, or moral convictions should be included in the model separately.

Further adjustments to the models pertain to the factors influencing social identity. The consistent predictive factors remain similar to those of the initial models. Injustice perceptions (as an overarching term for issue appraisals) seemed to be required for individuals to identify with an activist group. This is a strong influence on both social identity and collective action. It is also inherently connected to moral beliefs and norms. Efficacy needs to be included in an altered form. It should be seen as interchangeable with a sense of agency. Collective efficacy beliefs and a sense of agency can play the same role, as they give individuals a sense of action. One aims to achieve something specific, while the other focuses more on the attempt to do something. Nevertheless, there needs to be a wish to act on the issue, which is represented in this interchangeable factor. Both the injustice perceptions and the collective efficacy/sense of agency have a direct and indirect influence on collective action. They motivate action generally, as an individual perceives an issue as unjust and wants to act to feel like they are doing something against said issue. These motivations also lead to social identity, which facilitates the collective action and allows for social connections that share similar motivations. Social connections, group identification, group dynamics, participatory efficacy, self-identity (support of and development of), and image are all factors that can have an influence on social identity and/ or collective efficacy but are not universal. They could be included in a model as possible effects, as they can have strong consequences.

The discussed adjustments are visualized in Figure 6.8. This is merely a suggestion of possible relations and requires further empirical testing. The model mainly serves the purpose of visualizing the findings and discussed relations. It should be noted that the interconnections as mentioned throughout the text are numerous and thus were not all possible to include in a simple model. The twin core included in this model serves to represent the close connection between the two factors and the reinforcement of the norms and morals through the social identity. Futhermore, it indicates normative fit categorization. The *Possible Additional Factors* represent the factors mentioned above, which can have a strong influence but could also be irrelevant to individuals. This is argued to be largely dependent on the individuals' valuation of those factors.

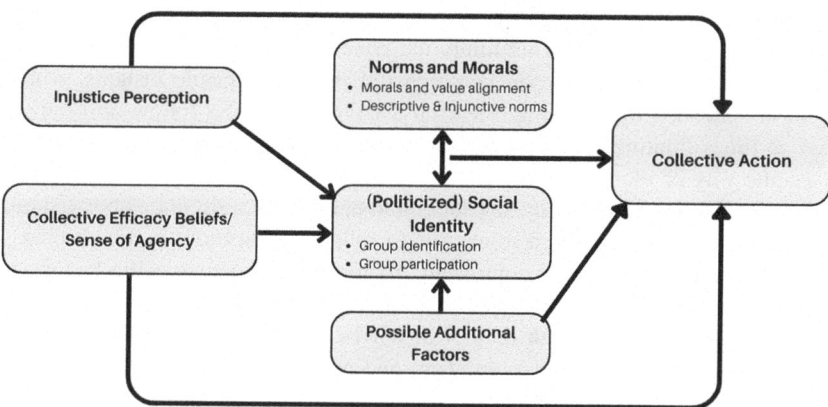

Figure 6.8 Suggested adjusted model of collective environmental action

6.4   Limitations, Critical Reflections, and Future Research

While this thesis opens up a critical realm of collective action research concerning environmental issues, it is limited in the conclusions that can be drawn from the empirical findings. The limitations are due to both methodological factors, as well as the reach and nature of the research. They should be considered in the discussion of the findings to contextualize and put into perspective the results. Furthermore, they illustrate future research opportunities to strengthen, extend and support the findings.

The limitations of the methods can be reiterated after conducting the analysis. Both the narrative interviews and the coding method have their drawbacks, which could be seen throughout the research. The narration opportunity brought about new factors and initial perspectives, but it varied in its functionality. The extent of the narrations varied from person to person. Some individuals brought up many points that influenced their decision, whereas others gave only a short response. This narration could be made more challenging because the decision-making process is not a concrete event to retell. Nevertheless, the point of the narrative interviews is to give individuals the opportunity to bring up what they find important. This had varying successes and provided two main challenges. On the one hand, the short narrations forced some interviews to run mainly off the prompts, negating the benefits of the narration. On the other hand, when

new factors were brought up, these could not be addressed with all the interviewees (e.g., self-image). This limits the conclusions that can be drawn from
these new influences. The open narration did provide valuable insights, which
could be extended by executing them on a larger scale. The follow-up prompts
were at times limiting, as they could push individuals to comment on a topic in
a specific way or mention an influence that they did not perceive as significant.
This is one of the reasons why several emotional factors were not elaborated on,
as they were often deliberated solely due to follow-up questions and with phrasing that would suggest some construing. To investigate more specific factors, it
might prove useful to conduct more structured interviews or even quantitative
evaluation of surveys (though these will not be able to encompass all possible
factors). Furthermore, different empirical and analytical methods could be used
to further extend the understanding of the processes that lead individuals to join
environmental activism groups. From expert interviews to media analyses, there
is a plethora of opportunities to explore.

The analysis through coding brings about two main limitations. First, as the
code was based on existing literature, the bias of previous research was introduced. The theoretical background can provide a helpful structure, but the biases
that form by using it to analyze findings still need to be recognized. Second,
analyzing individual codes creates the challenge of a loss of context. Offsetting
this lack of context by placing an additional focus on individual stories allowed
the factors to be put back into some context, although some loss is unavoidable.

The current research is an initial insight into a local activist group and the
motivations of its members, as well as non-member activists. This provides a
real-world perspective of possible influences and patterns in decision-making processes to join a group like XR. Nevertheless, as the research is quite specific, the
transferability is limited, as the focus was rather narrow. More research into motivations for environmental activism on a national and international level would be
necessary to strengthen the findings on a larger scale. Furthermore, research into
the diversity issues mentioned in the *Background* chapter is necessary, as interviewees that were predominantly white prevailed in the current study. The results
bring up crucial points of consideration and it is made apparent that the existent
collective action models require adjustment in the environmental realm. Accord-
ingly, future research should elaborate on possible factors and specify empirical
research to further investigate emotional effects, self-identity effects, and social
connection effects. These factors emerging from the research seemed to leave
the most room for future research, though research into the interactions between
all factors and the application of newly suggested relations for various groups
and locations would also be of great benefit. Additionally, two further points

could provide valuable insights. The efficacy belief and sense of agency effect could benefit from closer investigation into what determines the weighting of the two and how individuals perceive their groups as efficacious. Furthermore, the valuation of different factors could increase understanding of individual decision-making and how this differs from person to person. This research area remains extremely relevant and provides promising future research.

Conclusion

7

This thesis utilized social identity models on collective action to initiate an investigation into what motivates participation in environmental activism groups—specifically analyzing the case of the local XR group in Stuttgart. The environmental background of the group sets it apart fundamentally from other social action groups, due to the issue's structural-incidental dichotomy, its lack of a clear other, and its multifaceted nature. The relevance of addressing this instance of collective action is apparent because of the need in society to bring forward vital scientific issues through activism. As it is not previously investigated specifically or independently, exploring how applicable existing collective action models are on environmental issues is consequential. As has been shown, the existent models are not sufficient in portraying social identification processes central to environmental activism groups, and not all existent research on influencing factors is applicable. This study provides an initial insight into possible points of divergence from other collective action groups, by researching the decision-making processes to join an environmental activist group through observations and narrative interviews with XR and non-XR activists. The investigation is founded on the applicability of the models SIMCA, EMSICA, and SIMPEA, based on the SIA, and finding commonalities, as well as divergences from these models and further studied relationalities of relevant factors.

The research questions, inherently intertwined, were addressed with suggested patterns, model adjustments, and support of the centrality of social identity. The essential suggested adjustments are visualized in Figure 12. (Politicized) social identity and norms and morals are suggested as a twin core, as norms and morals are both a central factor of identification and can be reinforced through the social identity. They play a necessary role in which group individuals chose to identify with but can be subject to perceptions. Social identity and the social identity approach proved to be central due to the social nature of collective action, the

Y. Plate, *Social Identity Motivators in Environmental Collective Action*, BestMasters, https://doi.org/10.1007/978-3-658-44047-3_7

need for collaboration, and the interconnectedness of the issue to be addressed. The more radical form of activism presupposes politicization and requires previous training, planning, and preparation, which makes social identity and collective action close to synonymous. Factors such as belief in collective efficacy or a sense of agency, moral and norm alignment, and perceptions of injustice influence a motivation to join collective action, which is done through social identity. Other collective action addressing the issue is possible, but the type of collective action is primarily chosen through the social identity by aligning with their descriptive norms. Perception of injustice is seen as both an emotional factor and an issue appraisal factor, encompassing this feeling of injustice, as well as moral convictions and environmental threat perceptions. It can be seen as a prerequisite to initiating action in an individual. The issue needs to be seen as such and as necessary to fight against, for an individual to seek out collective action groups.

The influences on identification with a specific politicized social group and strength thereof can differ depending on the size, radicalness, and structuring of the group, as well as on the individuals' valuation of different factors. This could already be seen in the comparison of FFF and XR. The variation between groups opened up for consideration the possible interchangeability of efficacy and a sense of agency. These two factors seemed to play a similar role in allowing the individuals to feel engaged in action through social identification. While XR participants valued a sense of agency over perceived efficacy beliefs, FFF members usually strongly valued the collective efficacy they perceived the group to have. Therefore, the interchangeability could relate to outside factors like personal valuation or size of the group.

The three main factors norms and morals, injustice perceptions, and collective efficacy retain their significance but require some adjustment in their meaning and positioning to be applicable to environmental action groups. This can be seen in Figure 12, where new relations as supported by the current research are proposed. Additional factors were discovered to have strong yet not universal effects and require further investigation. Nevertheless, they bring forward noteworthy results. Social connections to people in an activist group can be determining in an individual's decision-making progress. The social connections within a group can additionally strengthen social identity. Shared emotions and emotions alleviated through group participation provided insufficient results to draw conclusions from. However, since one of XR's foci was named to be processing emotions together (e.g., through resilience meet-ups), this could still remain relevant. While responsibility proved to play more of an interacting role, self-identity and image were relevant enough to act as determining factors for some individuals. The support or development of one's self-identity through social identification and activism could be seen in various forms. Image was similarly relevant, especially when shared amongst or originating from close social connections. Through the necessary adjustment and new possible determining factors, it is made apparent that the topic remains pertinent to explore.

As the present study is an initial insight into the possible divergences of environmental collective action groups in this topic, further research would be beneficial to further the investigation into this realm. Elaboration on additional factors, exploration of groups of different sizes and locations, and methodological diversity would be necessary to gain a coherent picture of the interactions between factors influencing environmental activism participation. The convergence of collective efficacy beliefs and a sense of agency, as well as the valuation of various factors, could be starting points for future research. Barriers presented through the unfulfilled factors could also provide insights into the recruitment of individuals into collective action groups. The unique issues the climate and biodiversity crisis present broaden the spectrum of collective action groups and introduces new dynamics into social identity research.

References

Amenta, E., Caren, N., Chiarello, E., & Su, Y. (2010). The Political Consequences of Social Movements. *Annual Review of Sociology, 36*(1), 287–307. https://doi.org/10.1146/ann urev-soc-070308-120029

Balthesen, E. (2019, November 15). What do we want? Climate Justice! *Klimareporter°*. https://www.klimareporter.de/international/what-do-we-want-climate-justice

Bamberg, S., Rees, J., & Seebauer, S. (2015). Collective climate action: Determinants of participation intention in community-based pro-environmental initiatives. *Journal of Environmental Psychology, 43*, 155–165. https://doi.org/10.1016/j.jenvp.2015.06.006

Bell, K., & Bevan, G. (2021). Beyond inclusion? Perceptions of the extent to which Extinction Rebellion speaks to, and for, Black, Asian and Minority Ethnic (BAME) and working-class communities. *Local Environment, 26*(10), 1205–1220. https://doi.org/10.1080/13549839.2021.1970728

Bliuc, A.-M., McGarty, C., Reynolds, K., & Muntele, D. (2007). Opinion-based group membership as a predictor of commitment to political action. *European Journal of Social Psychology, 37*(1), 19–32. https://doi.org/10.1002/ejsp.334

Bührle, H., & Kimmerle, J. (2021). Psychological Determinants of Collective Action for Climate Justice: Insights from Semi-Structured Interviews and Content Analysis. *Frontiers in Psychology, 12*, 695365. https://doi.org/10.3389/fpsyg.2021.695365

Dresing, T., & Pehl, T. (2015). *Praxisbuch Interview, Transkription & Analyse: Anleitung und Regelsysteme für qualitativ Forschende* (6th Edition).

Extinction Rebellion. (n.d.). *Extinction Rebellion*. Retrieved April 19, 2023, from https://rebellion.global/

Extinction Rebellion Germany. (n.d.). *XR Germany | Extinction Rebellion*. https://rebellion.global/groups/de-germany/#groups

Fahrion, G. (2019, August 19). Neue Umweltbewegung Extinction Rebellion: Greta Thunbergs radikale Geschwister. *DER SPIEGEL*. https://www.spiegel.de/wissenschaft/natur/extinction-rebellion-was-die-neuen-klima-aktivisten-planen-a-1282370.html

Fielding, K. S., McDonald, R., & Louis, W. R. (2008). Theory of planned behaviour, identity and intentions to engage in environmental activism. *Journal of Environmental Psychology, 28*(4), 318–326. https://doi.org/10.1016/j.jenvp.2008.03.003

Fisher, S. R. (2016). Life trajectories of youth committing to climate activism. *Environmental Education Research, 22*(2), 229–247. https://doi.org/10.1080/13504622.2015.1007337

Flick, U. (2014). Gütekriterien qualitativer Sozialforschung. In N. Baur & J. Blasius (Eds.), *Handbuch Methoden der empirischen Sozialforschung* (pp. 411–423). Springer VS.

Fritsche, I., Barth, M., Jugert, P., Masson, T., & Reese, G. (2018). A social identity model of pro-environmental action (SIMPEA). *Psychological Review, 125*(2), 245–269. https://doi.org/10.1037/rev0000090

Fritsche, I., & Masson, T. (2021). Collective climate action: When do people turn into collective environmental agents? *Current Opinion in Psychology, 42*, 114–119. https://doi.org/10.1016/j.copsyc.2021.05.001

Hardin, G. (1968). The Tragedy of the Commons. *Science, 162*(3859), 1243–1248.

Harding, S. G. (1986). *The science question in feminism* (1. Edition). Open University Press.

Haugestad, C. A., Skauge, A. D., Kunst, J. R., & Power, S. A. (2021). Why do youth participate in climate activism? A mixed-methods investigation of the #FridaysForFuture climate protests. *Journal of Environmental Psychology, 76*, 101647. https://doi.org/10.1016/j.jenvp.2021.101647

Hsieh, H.-F., & Shannon, S. E. (2005). Three approaches to qualitative content analysis. *Qualitative Health Research, 15*(9), 1277–1288. https://doi.org/10.1177/1049732305276687

IPCC. (2018). Summary for Policymakers. In IPCC (Ed.), *Global Warming of 1.5°C: An IPCC Special Report on the impacts of global warming of 1.5°C above pre-industrial levels and related global greenhouse gas emission pathways, in the context of strengthening the global response to the threat of climate change, sustainable development, and efforts to eradicate poverty* (pp. 3–24). Cambridge University Press. https://doi.org/10.1017/9781009157940.001

Jansma, A., van den Bos, K., & Graaf, B. A. de (2022). Unfairness in Society and Over Time: Understanding Possible Radicalization of People Protesting on Matters of Climate Change. *Frontiers in Psychology, 13*, 778894. https://doi.org/10.3389/fpsyg.2022.778894

Jugert, P., Greenaway, K. H., Barth, M., Büchner, R., Eisentraut, S., & Fritsche, I. (2016). Collective efficacy increases pro-environmental intentions through increasing self-efficacy. *Journal of Environmental Psychology, 48*, 12–23. https://doi.org/10.1016/j.jenvp.2016.08.003

Landmann, H., & Rohmann, A. (2020). Being moved by protest: Collective efficacy beliefs and injustice appraisals enhance collective action intentions for forest protection via positive and negative emotions. *Journal of Environmental Psychology, 71*, 101491. https://doi.org/10.1016/j.jenvp.2020.101491

Laville, S., Noor, P., & Walker, A. (2019, February 15). 'It is our future': children call time on climate inaction in UK. *The Guardian*. https://www.theguardian.com/world/2019/feb/15/children-climate-inaction-protests-uk

Mayring, P. (2000). Qualitative Content Analysis. *Forum: Qualitative Social Research* (2).

Mayring, P., & Fenzl, T. (2014). Qualitative Inhaltsanalyse. In N. Baur & J. Blasius (Eds.), *Handbuch Methoden der empirischen Sozialforschung* (pp. 543–556). Springer VS.

Morrow, S. L. (2005). Quality and trustworthiness in qualitative research in counseling psychology. *Journal of Counseling Psychology, 52*(2), 250–260. https://doi.org/10.1037/0022-0167.52.2.250

Rainsford, E., & Saunders, C. (2021). Young Climate Protesters' Mobilization Availability: Climate Marches and School Strikes Compared. *Frontiers in Political Science, 3*, Article 713340. https://doi.org/10.3389/fpos.2021.713340

Rawls, J. (1971). *A theory of justice* (Original Ed.). Belknap Press.

Rees, J. H., & Bamberg, S. (2014). Climate protection needs societal change: Determinants of intention to participate in collective climate action. *European Journal of Social Psychology*, *44*(5), 466–473. https://doi.org/10.1002/ejsp.2032

Reicher, S., Spears, R., & Haslam, S. A. (2010). The Social Identity Approach in Social Psychology. In M. Wetherell & C. T. Mohanty (Eds.), *The SAGE Handbook of Identities* (pp. 45–62). SAGE.

Rosenthal, G. (2014). Biographieforschung. In N. Baur & J. Blasius (Eds.), *Handbuch Methoden der empirischen Sozialforschung* (pp. 509–520). Springer VS.

Schlosberg, D., & Collins, L. B. (2014). From environmental to climate justice: climate change and the discourse of environmental justice. *WIREs Climate Change*, *5*(3), 359–374. https://doi.org/10.1002/wcc.275

Schmitt, M. T., Mackay, C. M., Droogendyk, L. M., & Payne, D. (2019). What predicts environmental activism? The roles of identification with nature and politicized environmental identity. *Journal of Environmental Psychology*, *61*, 20–29. https://doi.org/10.1016/j.jenvp.2018.11.003

Simon, B., & Klandermans, B. (2001). Politicized collective identity: A social psychological analysis. *American Psychologist*, *56*(4), 319–331. https://doi.org/10.1037/0003-066X.56.4.319

Stern, P. C. (2000). New Environmental Theories: Toward a Coherent Theory of Environmentally Significant Behavior. *Journal of Social Issues*, *56*(3), 407–424. https://doi.org/10.1111/0022-4537.00175

Stürmer, S., & Simon, B. (2004). Collective action: Towards a dual-pathway model. *European Review of Social Psychology*, *15*(1), 59–99. https://doi.org/10.1080/10463280340000117

Tajfel, H. (1978). *Differentiation between Social Groups: studies in the social psychology of intergroup relations*. Academic Press Inc.

Taylor, M. (2020, August 4). The evolution of Extinction Rebellion. *The Guardian*. https://www.theguardian.com/environment/2020/aug/04/evolution-of-extinction-rebellion-climate-emergency-protest-coronavirus-pandemic

Thackeray, S. J., Robinson, S. A., Smith, P., Bruno, R., Kirschbaum, M. U. F., Bernacchi, C., Byrne, M., Cheung, W., Cotrufo, M. F., Gienapp, P., Hartley, S., Janssens, I., Hefin Jones, T., Kobayashi, K., Luo, Y., Penuelas, J., Sage, R., Suggett, D. J., Way, D., & Long, S. (2020). Civil disobedience movements such as School Strike for the Climate are raising public awareness of the climate change emergency. *Global Change Biology*, *26*(3), 1042–1044. https://doi.org/10.1111/gcb.14978

Thierbach, C., & Petschick, G. (2014). Beobachtungen. In N. Baur & J. Blasius (Eds.), *Handbuch Methoden der empirischen Sozialforschung* (pp. 855–866). Springer VS.

Thomas, E. F., Mavor, K. I., & McGarty, C. (2012). Social identities facilitate and encapsulate action-relevant constructs. *Group Processes & Intergroup Relations*, *15*(1), 75–88. https://doi.org/10.1177/1368430211413619

Thomas, E. F., McGarty, C., & Mavor, K. I. (2009). Aligning identities, emotions, and beliefs to create commitment to sustainable social and political action. *Personality and Social Psychology Review: An Official Journal of the Society for Personality and Social Psychology, Inc*, *13*(3), 194–218. https://doi.org/10.1177/1088868309341563

Turner, J. C., Hogg, M. A., Oakes, P. J., Reicher, S. D., & Wetherell, M. (1987). *Rediscovering the social group: A self-categorization theory*. Basil Blackwell.

van Zomeren, M. (2016). Building a Tower of Babel? Integrating Core Motivations and Features of Social Structure into the Political Psychology of Political Action. *Advances in Political Psychology, 37*, 87–114. https://doi.org/10.1111/pops.12322

van Zomeren, M., Kutlaca, M., & Turner-Zwinkels, F. (2018). Integrating who "we" are with what "we" (will not) stand for: A further extension of the Social Identity Model of Collective Action. *European Review of Social Psychology, 29*(1), 122–160. https://doi.org/10.1080/10463283.2018.1479347

van Zomeren, M., Postmes, T., & Spears, R. (2008). Toward an integrative Social Identity model of Collective Action: A quantitative research synthesis of three socio-psychological perspectives. *Psychological Bulletin, 134*, 504–535. https://doi.org/10.1037/0033-2909.134.4.504

van Zomeren, M., Postmes, T., & Spears, R. (2012). On conviction's collective consequences: Integrating moral conviction with the social identity model of collective action. *British Journal of Social Psychology, 51*, 52–71. https://doi.org/10.1111/j.2044-8309.2010.02000.x

Walker, G. (2020). Environmental Justice. In A. L. Kobayashi (Ed.), *Encyclopedia of human geography* (Vol. 4, pp. 221–225). Elsevier.

Weyler, R. (2018, January 5). *A Brief History of Environmentalism*. Greenpeace International. https://www.greenpeace.org/international/story/11658/a-brief-history-of-enviro nmentalism/

Xiang, P., Zhang, H., Geng, L., Zhou, K., & Wu, Y. (2019). Individualist-Collectivist Differences in Climate Change Inaction: The Role of Perceived Intractability. *Frontiers in Psychology, 10*, 187. https://doi.org/10.3389/fpsyg.2019.00187

GPSR Compliance

The European Union's (EU) General Product Safety Regulation (GPSR) is a set of rules that requires consumer products to be safe and our obligations to ensure this.

If you have any concerns about our products, you can contact us on ProductSafety@springernature.com

In case Publisher is established outside the EU, the EU authorized representative is:

Springer Nature Customer Service Center GmbH
Europaplatz 3
69115 Heidelberg, Germany